OFF-EARTH

OFF-EARTH

ETHICAL QUESTIONS AND QUANDARIES FOR LIVING IN OUTER SPACE

ERIKA NESVOLD

THE MIT PRESS
CAMBRIDGE, MASSACHUSETTS
LONDON, ENGLAND

The MIT Press would like to thank the anonymous peer reviewers who provided comments on drafts of this book. The generous work of academic experts is essential for establishing the authority and quality of our publications. We acknowledge with gratitude the contributions of these otherwise uncredited readers.

This book was set in Arnhem Pro and Frank New by New Best-set Typesetters Ltd. Printed and bound in the United States of America.

Library of Congress Cataloging-in-Publication Data

Names: Nesvold, Erika, author.
Title: Off-Earth : ethical questions and quandaries for living in outer space / Erika Nesvold.
Description: Cambridge, Massachusetts : The MIT Press, [2023] | Includes bibliographical references and index.
Identifiers: LCCN 2022006370 (print) | LCCN 2022006371 (ebook) | ISBN 9780262047548 (hardcover) | ISBN 9780262372527 (epub) | ISBN 9780262372534 (pdf)
Subjects: LCSH: Space colonies—Moral and ethical aspects. | Space colonies—Social aspects.
Classification: LCC TL795.7 .N47 2023 (print) | LCC TL795.7 (ebook) | DDC 629.44/2—dc23/eng/20221003
LC record available at https://lccn.loc.gov/2022006370
LC ebook record available at https://lccn.loc.gov/2022006371

10 9 8 7 6 5 4 3 2 1

CONTENTS

PREFACE

"We'll worry about that later" is not something I ever expected to hear from a leader in the space industry. And yet, sitting in a conference room in Mountain View, California, in 2016, that's exactly the response I received from the CEO of a (now-defunct) space mining company when I asked him how he planned to address the planetary protection risks of his lunar mining equipment. I was spending the summer in Silicon Valley as part of a NASA (National Aeronautics and Space Administration) research program that brought together astrophysicists, like myself, and machine learning experts to tackle big data problems in planetary defense. The program directors had lined up a number of visits with people working in the "NewSpace" industry: entrepreneurs and engineers who hoped to make space travel cheaper and easier while building a new and highly profitable economy in space. I've always been a proponent of not only space science and human space exploration but also the longer-term goal of permanent human settlement in space. At the time, I was also considering a career shift from research science to industry, so I was excited to talk to these industry experts to get a sense of what open questions in the field remained to be addressed.

To my dismay, I quickly discovered that the private space industry professionals I talked to that summer, even the ones who shared my dream of building communities in space, had little interest in questions that seemed crucial to me: How will we protect the space environment from our own activities? How will workers in space be shielded from exploitation? Who will be responsible for settling conflicts between individuals living in space? *We'll worry about that later* was the consistent response from the space industry representatives I encountered. Instead, they seemed to be focused exclusively on technical challenges like reusable rocket designs, economic strategies for making space activities financially feasible, and legal structures that would invigorate rather than inhibit their industry. This tunnel vision baffled me. Space settlement advocates often advertise space as a blank slate where we can build utopian societies free from the crowded territory and bloodied history of our terrestrial home. But adopting a "worry about it later" attitude toward human rights and ethics strikes me as a path to repeating the tragedies of that history through ignorance.

When I returned home to the East Coast from California and began researching some of these issues, I was relieved to discover that there are plenty of space ethicists and other scholars who are already considering how we can live justly and ethically in space. There also are countless researchers studying human society on Earth who have never considered their work in the context of space but have much to offer in the way of cautionary tales from our past and analyses of the origins of injustice and exploitation today. But there remains a disconnect between this wealth of knowledge and advice and the people who are actually working in the space industry.

During my research, I quickly identified one possible cause of this disconnect. As I read, I became uncomfortably aware of my own lack of education in history, sociology, anthropology, ethics, and the other humanities fields that were missing from my STEM (science, technology, engineering, and math) education. Given that most of the workers in the space industry, both public and private, are trained in technology, engineering, and physics rather than the humanities, perhaps it simply doesn't occur to them that the humanities can help us build better space communities. So, I reached out to experts in those fields and interviewed them about their research with an eye toward the implications for our future life in space. The result was a thirteen-episode podcast called *Making New Worlds*.[1] My goal for the podcast was never to determine or dictate exactly which choices we should be making to ensure the best possible lives for future generations living in space. Instead, I hoped to prompt space settlement enthusiasts to join me in my search for the most important questions we should be asking in these early days of humanity's presence in space, and to recognize the abundance of accumulated knowledge that is already available to us, if we can manage to look beyond our technical training and default worldview.

This book is an expansion of the podcast, featuring additional interviews with relevant experts, further examples of historical case studies that can guide our future community-building, and updates on the progress of the private space industry, which has been busy in the years since I created *Making New Worlds*.

It's unlikely that I'll ever get to set foot on a space settlement, but all of us, even the majority who are destined to never leave our home planet, can work together to build a better future for our spaceborn descendants who will spend their lives there. And the time to start that work is now.

INTRODUCTION

In 1869, Unitarian minister Edward Everett Hale published a novella describing a hollow artificial satellite, "The Brick Moon," cheerfully populated by men, women, and children who thrived on its surface after they were accidentally launched into orbit. Hale's story was one of the first known works of fiction to describe humans permanently (if unintentionally) living in space, published almost thirty-five years before humans even invented the airplane. The same year the Wright brothers made their first flight, physicist Konstantin Tsiolkovsky published an equation for calculating the amount of fuel needed for a given spacecraft launch or maneuver, which eventually become known as Tsiolkovsky's rocket equation (although it was independently derived by mathematician William Moore nearly a century earlier). Tsiolkovsky didn't live to see the first rocket launched into space, but he theorized that humans would one day venture throughout the Milky Way galaxy.

Throughout humanity's history of scientific innovation, our imagination has often raced ahead of our technological progress, especially in space. *Star Trek*, the television show that inspired so many of today's space workers, premiered nearly three years before the Apollo 11 mission left for its three-day trip to the Moon. By the time Armstrong and Aldrin set foot on the lunar surface, many of the viewers watching breathlessly on Earth were convinced that humanity would establish permanent outposts on the Moon and even Mars by the end of the twentieth century. Fifty years later, the most we've managed is the International Space Station (ISS), a small outpost a few hours' journey from the surface of the Earth, inhabited by half a dozen temporary staff. The ISS is a technological marvel, to be sure, and potentially the most expensive construction project in human history, but it is far from the self-sustaining colonies envisaged by science fiction authors and space enthusiasts of the twentieth century. But while our movement outward into space has been slower than expected thanks to political, technological, and financial limitations, a new generation of space settlement advocates is emerging.

Raised on a steady diet of *Star Trek* and American exceptionalism, this generation is coming to power on the crest of a different technological revolution: the digital era. The stories we've been told since birth, and those we continue to tell each other, promise a bright future for anyone bold enough to step forward and take a risk. These start with stories about the past, like "The western American frontier was a place of limitless opportunity for brave and hard-working settlers," and continue with stories about the present, like "Bill Gates built a computer in his garage and became one of the richest men in the world," and culminate with stories about the future, like "Innovation and courage will let us build utopias in space while becoming fabulously wealthy at the same time."

Stories are how we learn from our past, interpret our present, and plan our future. It's long past time for this new space generation to listen to the other stories, the ones that aren't told as loudly or officially, like "The westward expansion of the American nation devastated entire cultures, destroyed landscapes, and exterminated species," or, "The same technologies and industries that put the world's knowledge at our fingertips have also worsened the global wealth gap and incentivized labor exploitation," or, "What if our future in space ends up more like the war-torn *Star Wars* than the utopian *Star Trek*?"

Exploration, migration, and expansion are common themes in human history, and many space settlement proponents point to our species' spread across the globe as evidence that settling space is a natural next step in our story. After all, history has demonstrated that migration can lead to innovation, increased wealth (for some), and the opportunity to try new ideas about the best ways to form a society. But it has also meant empire-building, invasion, environmental devastation, and war. We need to ask ourselves: what part of our global history will we choose to bring with us into space, and what will we leave behind?

In 1977, physicist Gerard K. O'Neill published one of the first serious roadmaps for space settlement, titled *The High Frontier*, which provided detailed plans for building rotating space stations in high Earth orbit. When writer and futurist Stewart Brand published O'Neill's early work in his counterculture magazine *CoEvolution Quarterly*, Brand's friend Wendell Berry, a poet and regular contributor to the magazine, reacted with scathing criticism of the plan. "Much is made of the fact that the planners' studies 'continue to survive technical review,'" Berry wrote. "But there is no human abomination that has not, or could not have, survived technical review." Berry went on to list examples of the horrors that humans have committed against the environment and each other, pointing out that "all have been technically feasible, and usually economically feasible as well."[1]

To avoid repeating the mistakes of our past, we need to figure out the kind of world we want future space settlers to inhabit. This means determining not only the lift capacity of the rockets that will carry them there and the design of their habitats when they arrive but also the ways that they'll live with each other, interact with their environments, and embody the values that we pass down to them. After all, we want our descendants in space to thrive, not just survive. In fact, we want *everyone* to flourish, in space and back on Earth, to have their physical needs met in a safe environment, to enjoy abundant opportunities for growth, education, and social connection, to practice their cultures and religions freely, and to provide the same for the next generation. Our plans for space should be able to pass an ethical review not just a technical one, but more than that, they should reflect our values. They should represent our demand for a better society than the one we were born into, one that deserves to spread beyond the planet where we evolved.

To perform a technical review, the head of a company that manufactures rockets, satellites, or space habitats would recruit scientists and engineers who can comment with expertise on their designs and prototypes. To perform an ethical review, space settlement organizations should similarly recruit experts in fields such as sociology, history, philosophy, and ethics. This book is intended to help bridge the gap between these experts and the workers and decision makers in today's space industry. It is specifically aimed toward readers who care about humanity's future in space but have minimal training and experience in sociological analysis or ethical frameworks.

Of course, you don't need an advanced degree in rocket science *or* the social sciences to help shape the plans for humanity's future in space. That future belongs to all of us. Space agencies depend on taxes and public support to continue their work. The private space industry is still fragile and new, and it also relies on public goodwill. Perhaps you are working or studying in one of the fields that will contribute to space settlement, like aerospace engineering, space medicine, mining, or ecology. If you have children, your descendants may one day inhabit communities built on other worlds.

This book is written with those future generations in mind, to recenter our conversations about space around the people we're trying to build a better future for, both in space and on Earth.

It's easy to get lost in the grand scale of space settlement when we're talking about time scales far longer than our life spans and distances far greater than the longest journey on Earth. As you ponder these Big Questions about "the future of humanity," try shifting your perspective. Picture yourself as one of the settlers. What does your day-to-day life look like? Will you have a job you enjoy? Will you worry about how to put food on the table? What will your kids do all day? What will you miss about Earth: fresh fruit, thunderstorms, a clear view of the sky? What will you dream about at night? The following chapters will introduce you to the larger decisions and challenges that we'll face as a society in space, but every one of those decisions will impact the lives of individuals, just as the choices that humans have made in the past have affected regular people throughout history who were simply living their lives. Each chapter will begin with a handful of fictional vignettes to help you imagine these experiences at a personal level. Keep that perspective in mind: Yes, this book is about questions that will determine the fate of our species in space, but it's also about you, and the kind of world you want to build there.

I

HOW DO WE BEGIN?

1

SHOULD WE SETTLE SPACE?

The year is 1635 CE, and you are grappling with a difficult choice: uproot your family for a dangerous voyage across the sea to seek refuge in the New World, or risk further persecution for your beliefs here at home. Your minister says that God will provide in the land of freedom, where you will all safely build His true church. But how can you be sure that the life you will have there, in the wilderness of a strange land, will be any better than the one you have here?

The year is 2017 CE, and you don't know what to do. Your son has refused to carry drugs for the local gang, and now your neighbor tells you that there is a price on his head. If you send your son away, you may never see him again, but he's only fourteen, and you cannot hope to protect him here. You have a cousin you've never met, living in a country you've never seen, who may be able to take him in—if you can get him there. Should you risk his arrest and deportation, or even his death in the desert, to try to send him north across the border?

The year is 2100 CE, and the day has finally come. Applications for the first space settlement have opened. You've been thinking through the possibilities, discussing the pros and cons with your family and friends, but now it's up to you. Will you gamble everything you have on the chance to be a part of the biggest step forward in human history? You don't have a bad life here on Earth. In fact, you think you could really make a difference here, do some good for the world. But is that really all you're meant for? Maybe things could be better in space. Maybe *you* could be better in space. Or maybe you'd be miserable, homesick, and trapped in a harsh life of deprivation with no way back. What will you choose?

——————

MOTIVATION

WISDOM FROM *JURASSIC PARK*

Private spaceflight companies face a number of as yet unanswered questions about the feasibility of their industry: Can we design rockets that are cheap enough to get an entire space mining or space tourism industry up and running? Can we build habitats in space that can adequately protect human residents from the deadly effects of radiation outside Earth's atmosphere? In the longer term, people who want to see humans settle space permanently will face further obstacles: Can healthy human fetuses develop in the low-gravity, high-radiation environment of space? Can we develop and maintain stable ecosystems away from Earth?

These questions are crucial for determining whether space settlement is even possible for our species. But as we work through them, we should heed the words of the shrewd, if fictional, Dr. Ian Malcolm of *Jurassic Park*: "Your scientists were so preoccupied with whether or not they could that they didn't stop to think if they should." Let's follow Dr. Malcolm's advice by asking ourselves: *Should* we settle space? Are we ready, or are we morally obligated to wait until we've evolved, culturally or physically, into a species better able to handle the rigors of off-world life? Should we attempt to spread our civilization beyond our home planet at all?

Space travel is not an endeavor without cost. The debate about NASA's budget springs up every couple of years, and private spaceflight companies are certainly acutely aware of the expense of rocket fuel. There's also the opportunity cost: what else could we be spending our nations' budgets and scientists' brainpower on? And, as we are reminded suddenly and usually explosively every couple of decades, there is a human cost to being a spacefaring species. Space is dangerous, and sometimes people die. Most recently, as of this writing, the seven crew members of the space shuttle *Columbia* were killed when the shuttle broke up during re-entry into Earth's atmosphere in 2003. The Columbia Accident Investigation Board asserted, as paraphrased by NASA administrator Mike Griffin, that if US spaceflight was to continue despite the danger and expense, "the goals ought to be worthy of the cost and the risk and the difficulty of the enterprise."[1]

What goals are worthy of the cost and risk of not only traveling to space but attempting to live there permanently? Advocates for space settlement have written countless books, articles, and speeches outlining the reasons that we must expand our civilization into space. These goals range from the practical (to ensure the continuation of the species in the event of a catastrophe) to the spiritual (to achieve our destiny), but a common thread through all of them is the idea that space settlement is a global endeavor, not confined to any single nation or company.

Unlike the 1960s, when the US rallied to support the Space Race against the USSR to ensure that an American would be the first to set foot on the Moon, space settlement advocates today usually frame their rhetoric to speak on behalf of all humanity, or even all life on Earth. And yet only a small fraction of Earth's population has been engaged in this debate. In fact, the space settlement community today is overwhelmingly male, white, and Western. This makes it likely that their arguments are biased by their own shared culture and background, and not as universal as they might assume.

Let's take a broad look at the arguments that space settlement advocates use today, along with counterarguments raised by philosophers and ethicists, and ask ourselves how universal this debate really is and what it reveals about the unexamined values of the debaters.

SPACE AS DESTINY

OUR FUTURE LIES IN THE STARS

In the first pages of his book, *The Case for Mars*, Mars Society founder Robert Zubrin quotes renowned science fiction author Arthur C. Clarke, who warns that, should we fail to meet the challenge of space, we will have "turned [our] back upon the still untrodden heights and will be descending again the long slope that stretches, across a thousand million years of time, down to the shores of the primeval sea." Later in the same book, Zubrin argues that failing to settle space (and, in particular, to terraform Mars) would constitute *"failure to live up to our human nature and a betrayal of our responsibility as members of the community of life itself"* (emphasis his).[2]

The belief in space as our destiny is one of the older arguments for space settlement. Konstantin Tsiolkovsky, one of the discoverers of the rocket equation, was motivated by the belief that rocket technology would one day allow human civilization to spread throughout the galaxy. Revered as one of the fathers of rocket science, Tsiolkovsky passed more than equations down to his intellectual descendants in the Soviet and Russian space programs. He also perpetuated the philosophy of Russian cosmism, a movement founded in the mid-nineteenth century by one of Tsiolkovsky's teachers, Nikolai Fyodorov, who believed that space travel would be necessary to both help humanity conquer death, and to house the large population of immortal humans that would result. A blend of science and mysticism, Russian cosmism includes a view of space as both a destination to be conquered and a tool to help us reach a higher form of existence. Tsiolkovsky himself believed not only that colonizing the galaxy was our destiny but that building a civilization in space would allow humanity to reach a state of perfection. One of Tsiolkovsky's most famous quotes, a favorite of the space exploration and settlement community, is usually translated as, "The Earth is the cradle of humanity, but one cannot live in the cradle forever."[3]

Cosmist ideals and values permeated the Russian space industry as well as Russian popular culture, while a parallel mythos of Manifest Destiny flourished in the US space program.[4] Originally coined by American journalist John O'Sullivan in 1845, "Manifest Destiny" refers to the belief that the North American continent was provided to the American settlers by divine providence, and that it was their God-given right and responsibility to expand across the country and spread their democratic civilization. This rhetoric has often been echoed in the narrative around space exploration and settlement, where space is framed as a blank slate, destined to be conquered and filled up with our well-structured new societies. These conversations about space rarely cite God or Christianity directly, but, ultimately, the thread of this argument can be traced back to

Christian dominion theology, an ideology based on the biblical command to "be fruitful, and multiply, and replenish the earth, and subdue it: and have dominion . . . over every living thing that moveth upon the earth."[5]

"It's important to think about the way that Christianity, as a source of culture, has influenced the West," anthropologist Deana Weibel said.[6] She studies the intersection of space and religion at Grand Valley State University. "Even if people aren't Christian, it's still baked into everything around them." Without the outside perspective provided by education in the social sciences and history, it can be difficult for someone to spot the biases or unique aspects of the culture they've spent their whole life inside. Many space advocates who use "destiny" arguments may be unaware that they are not citing universal human ideals, but merely those of hegemonic Western, heavily Christian cultures.

Even if we set aside the historical and cultural influences of Christianity, Weibel noted that the "destiny" argument for space settlement can take on a religious tone. She recalled interviewing a space worker involved with NASA's Human Research Program, a self-described atheist who "believed that it was absolutely true that humanity was going to get into space. 'We absolutely are,' he said. 'There's no conceivable way we're not going to settle in space.' And that kind of thing," Weibel noted, "it's not rational, it is completely based in faith."

Weibel cited the work of space historians who have argued that enthusiasm for space is itself a religion.[7] She described one of the most direct parallels between the two, that of a group identity: "That sense that, if you're into space, and you meet somebody who is also into space, you already have that immediate connection. And they understand you, and you have some shared values. And if you meet somebody who's not into space at all, it's kind of frustrating, because they don't get why it's cool . . . That idea that you're part of a group, and you can convert people into the group."[8]

That's what space settlement advocates are trying to do: to convert non-believers into the group. But an important step is to recognize that people outside the group are not simply undereducated in the benefits of space exploration; they may have different fundamental values and priorities. Carl Sagan, when asked about the intangible arguments for space travel, such as claiming our destiny in space, acknowledged that,

"This is fundamentally a religious argument, and not everybody shares that particular faith . . . If you had children who didn't have enough to eat, the idea of spending one hundred billion dollars or even three hundred or five hundred billion dollars to send some people to Mars would seem ludicrous."[9]

A similar argument to the "space as destiny" framing, but one with a more scientific tone, is the idea that space settlement is an expression of basic human nature. This argument is based on the belief that humans have an innate desire to explore, one that inevitably leads us into space. Carl Sagan made this point in his television show *Cosmos*: "Exploration is in our nature. We began as wanderers, and we are wanderers still. We have lingered long enough on the shores of the cosmic ocean. We are ready at last to set sail for the stars."[10]

Clearly this perspective is also not immune to poetry. But unlike the appeal to a spiritual destiny in space, advocates of the "exploration is human nature" argument cite scientific evidence that humans have evolved an urge to "wander," culturally, psychologically, or even genetically. For example, research indicates that a variant of a gene called DRD4 is associated with "novelty-seeking" behavior, a heritable tendency toward risk-taking that can stimulate exploration.[11] The presence of this gene appears to be correlated with early human migration on Earth: the farther a population spread across the planet, the higher the occurrence of this variant.[12] But as with many attempts to connect specific genes to general human behavior, this story stands on shaky ground. Does a higher incidence of the novelty-seeking variant *cause* people to migrate away from home as they seek new experiences? Or did it merely provide beneficial traits for those who did migrate, and was thus selected for after migration occurred?

Even if it's true that the urge to explore and migrate has a genetic component, this wouldn't imply that the human species as a whole is genetically destined to migrate to space. As a species, we are not nomadic (although nomadic societies still exist on Earth): plenty of humans have lived and died in the regions where they were born, even as other members of their societies struck out for new lands. To argue for space settlement as a parallel to historical human migration, the most one could say is that it is the nature of *some* humans to want to settle space. And why should this argument sway the rest of humanity, the significant majority of whom may prefer to remain behind on Earth?

Another wrinkle in the genetic argument is that the existence of a "wanderer" gene would only indicate that the human species is *likely* to wander into space, not that we necessarily *should*. The position that we should settle space because we are genetically predisposed to do so runs afoul of a logical fallacy often referred to as the "appeal to nature," which asserts that because something is "natural," it is therefore the good or right choice. Humans have a number of biological drives that can lead them to commit acts that human society generally regards as morally wrong. I don't mean to compare settling space to aggression, violence, or any of our other less-than-pleasant tendencies, just to point out that we should judge the decision to settle space on its own merits, rather than conceding to some kind of genetic inevitability.

SPACE AS PROGRESS

IF WE DON'T GROW, WE STAGNATE

A more common category of argument for space settlement suggests that while expansion into space is not necessarily inevitable, it is nevertheless beneficial for humanity. This category is much more practical, and includes, for example, pointing out specific resources in space that we lack here on Earth. There are certainly industries and consumers that would benefit from access to space resources like precious metals in asteroids or a microgravity environment for manufacturing. Space also provides opportunities that could benefit humanity as a whole, such as space-based solar power or physical space itself to relieve overcrowding.

But another class of arguments suggests that it is the act itself of expanding our civilization into space that would benefit humankind. The most practical example is technological: yes, there are precious metals to be mined in asteroids, but to mine them we would also need to develop better mining technologies, improved remote surveying and detection equipment, a reliable communications network, and food storage, life support, and medical technologies to keep the miners alive and healthy. All these technological advances would likely provide benefits for the non-spacefaring humans on Earth. On a much larger scale, advocates for terraforming other planets often argue that developing terraforming technology would vastly increase our understanding of Earth's climate system and provide us with tools for fighting climate change. On the other hand, space settlement skeptics wonder why we don't just prioritize developing these technologies directly, instead of gambling on tangential technologies.

Thanks to their isolation from Earth-bound societies and each other, space settlements will also provide an opportunity for what physicist Gerard O'Neill called "experiments in society-building." While he admitted that some of these experiments would likely fail, others "may succeed, and those independent social laboratories may teach us more about how people can best live together than we can ever learn on Earth, where high technology must go hand-in-hand with the rigidity of large-scale human groupings."[13]

Many space settlement advocates claim that settlement is not just *beneficial* for human society as a stimulus for innovation but necessary. They argue that expansion into space represents progress and that progress is required to maintain a healthy society. This is, once again, based on an assumption about human nature. Carl Sagan claimed that our species is the kind that "needs a frontier—for biological reasons. Every time humanity stretches itself and turns a new corner, it receives a jolt of productive vitality that can carry it for centuries."[14] This reference to space as a "frontier" is extremely common in discussions of space settlement and often explicitly references the western frontier of European settlement in North America. For example, O'Neill's famous proposal for space colonization via rotating stations in high Earth orbit, quoted in the previous paragraph, was titled *The High Frontier*. The National Space Society hopes to emulate "the opening of 'the New World' to western civilization,'" and attributes to this historical frontier "the development of the 'open society' founded on the principles of individual rights and freedoms."[15]

Other advocates have pointed to the darker side of this frontier argument: the claim that without a frontier to progress toward, society will stagnate. Jeff Bezos cites this as his personal motivation for directing his financial resources toward space settlement: "A life of stasis would be population control, combined with energy rationing. That is the stasis world that you live in if you stay . . . That, to me, doesn't sound like a very exciting world for our grandchildren's grandchildren to live in."[16] Robert Zubrin is a particularly strong proponent of this argument, warning that "without a frontier to grow in, not only American society, but the entire global civilization based upon values of humanism, science, freedom, and progress will ultimately die."[17] Zubrin is especially concerned with what he sees as evidence of "an ever more apparent loss of vigor of our society," including, among other things, "the banalization of popular culture."

In *The Case for Mars*, Zubrin explicitly cites nineteenth-century American historian Frederick Jackson Turner, who argued in 1893 that the moving frontier line between settled American territory and the unsettled "wilderness" inhabited by Indigenous people had indelibly shaped the character of both individual Americans living on the frontier and America as a nation. Turner's frontier thesis, along with the mythologized version of the western frontier in popular culture, significantly influenced American values and culture in the twentieth century and was taught in US history classes for decades. The theory is no longer accepted by most modern historians, but the rhetoric lingers, and is still evident as a motivator for space advocates and science fiction fans alike. Every episode of the original *Star Trek* series, as well as its first spinoff, *Star Trek: The Next Generation*, began by orienting the viewer to both the setting and the theme: "Space: the final frontier . . ."

The harsh living conditions in early space settlements may stimulate technological and societal innovation out of necessity. But they are just as likely (perhaps even more likely) to restrict opportunities for discovery and advancement. As the residents of these settlements struggle to survive and establish sustainable environments, there won't be much free time for basic scientific research or healthy social experimentation. One of Zubrin's claims in *The Case for Mars* is that space represents the ultimate opportunity for lasting equality and peace, because "only in a universe of unlimited resources can all men be brothers." But initially, and for a long time into the future, space settlements *won't* have unlimited resources. In fact, they'll have significantly fewer resources than they would on Earth. And those resources—down to the air they breathe—will be controlled by a handful of people, opening the population up to terrifying new methods of tyrannical behavior. As philosopher James Schwartz points out, "space settlements are much more likely to strain democratic principles than they are to encourage freedoms we usually associate with liberal democratic society."[18]

Science writer Martin Robbins critically examined what he dubbed the "pernicious myth" that space will inevitably improve human society: "You can sum it up like this: 'When we go into space, we will all magically become nice.' . . . Except every available single scrap of historical experience tells us that this is an incredibly naive and dangerous assumption to make."[19] Indeed, space advocates like Zubrin who use Turner's frontier thesis to claim that the space "frontier" will usher in a better society are making the rookie mistake of using a single data point (the European colonization of North America) to make their case. To make matters worse, they are viewing that single data point through what Schwartz points out are rose-colored glasses: an undereducated view of American history that only considers the positive results of the frontier for those (white, land-owning men) who benefitted from it, while neglecting the misery and oppression inflicted on everyone else.

More broadly, this conception of space settlement evokes the idea of progress as a moral imperative: that we must, at all times, continue to grow and expand as a society, because to do otherwise is to fail, to stagnate, to turn away from Clarke's "untrodden heights" and descend back toward our "primeval" origins. This veneration of progress is so ingrained in American culture that my American readers may, at this point, be wondering what could possibly be wrong with such self-evident statements as "Progress is good for society" and "Failure to progress is bad." But this belief, enhanced by America's obsession with the frontier thesis and Manifest Destiny, is far from universal. Social scientist Linda Billings, who has expended a significant amount of effort critiquing "the American rhetoric of manifest destiny," points out that non-Americans are often "baffled, if not offended, by this rhetoric."[20]

What are the alternatives to an ever-growing society centered around progress? The hypothetical static, frontier-less society that Zubrin and Bezos decry could also be called a steady-state, sustainable society. Perhaps there is an option between always-growing and always-declining, one in which we balance our physical needs with our resources while our intellectual and cultural evolution continue. In rejecting such a steady-state system, Zubrin and Bezos also reject the need to learn how to live sustainably within our environment.

Historian J. B. Bury proposed an oft-quoted definition of humanity's progress as civilization's movement "in a desirable direction," but he immediately pointed out that to measure progress, we have to know (and presumably agree upon) what our destination is and be able to tell when we are moving in that direction, a task that he calls "impossible."[21] What is the end goal of progress in a spaceward direction? Is the ideal society one in which we expand outward forever, never losing that all-important frontier? Is it one in which we learn how to live in harmony with our environment and each other? One where everyone's needs are met and people are free to explore the frontiers of art and science without worrying about survival? There is no single answer to these questions; utopia looks different to everyone.

SPACE AS SURVIVAL

WE SETTLE SPACE, OR WE DIE

There is an even more practical argument in favor of space settlement: the survival of the human species. Disaster movies and real-world scientists alike have offered us a plethora of plausible (and less plausible) scenarios in which the Earth might become uninhabitable for humans. We could be struck by a sudden, planetary-scale natural disaster like an asteroid impact, or a human-made, preventable (but no less deadly) disaster like global nuclear war. Today's leading candidate for the end of civilization is a slower, self-inflicted demise: the climate change–induced collapse of the food supply. Even if we have the fortitude, innovation, and luck to avoid these and all other potential disasters, the descendants of human-kind will still have to face the evolution of the Sun into a red giant. In one billion years, the increasing energy output of the Sun will evaporate all of Earth's oceans. The dry surface of the Earth will continue to heat until it becomes molten. Eventually, within eight billion years, the Sun will have expanded enough to engulf the Earth.[22] No matter what we do, the Earth will one day be a place where humans cannot survive.

This inevitability is frequently referenced by advocates of space settlement. In 2014, former astronaut Charles F. Bolden, Jr., at the time the NASA Administrator, summarized this position during a speech at a conference on human exploration of Mars: "If this species is to survive indefinitely, we need to become a multi-planet species."[23] Famed theoretical physicist Stephen Hawking made similar statements to the public in the years before his death, arguing that, "It is important for the human race to spread out into space for the survival of the species. Life on Earth is at the ever-increasing risk of being wiped out by a disaster, such as sudden global warming, nuclear war, a genetically engineered virus, or other dangers we have not yet thought of."[24] Elon Musk has frequently described the survival of the human race as being his primary motivation for settling space: "I do not have an immediate doomsday prophecy, but eventually, history suggests, there will be some doomsday event. The alternative is to become a space-bearing civilization and a multi-planetary species."[25]

Musk also sees this as a counterargument to the proposal that we should solve our problems on Earth first, before going into space, and to Carl Sagan's point that people struggling to feed their children might not rank space settlement as a high priority. Musk sees a "strong humanitarian argument" for ensuring the survival of the human species by establishing an auxiliary population in space, safe from a hypothetical global catastrophe. If such a catastrophe were to occur without a contingency in place, Musk insists, "being poor or having a disease would be irrelevant, because humanity would be extinct."[26] But in this scenario, in which the Earth is struck by catastrophe, leaving our space settlements the only refuge of humankind, our hypothetical poor or sick human is presumably still living on Earth, and is dead either way. Even if our imagined catastrophe is predicted far enough ahead of time that humans living on Earth could flee to space settlements, we would need an unimaginably advanced launch system and settlement capacity to be able to evacuate everyone. For the foreseeable future, space settlements will not represent salvation in the event of catastrophe, but rather something like the

Svalbard Global Seed Vault, which was not designed to prevent global crop failures, but rather as a kind of ark to safeguard against the total extinction of those crops.

Some environmentalists worry that the model of space as a refuge from the self-inflicted, deteriorating condition of the Earth could lead to a "disposable planet mentality": a path in which we use up each new environment we encounter, extracting all the useful resources we can and dumping our waste, then moving on to the next. The first environmental victim in this scenario would be the Earth, whose human-friendly ecosystem we are already well on our way to destroying. Will the allure of distant, "untouched" planets, still rich in valuable resources, pull our focus away from rehabilitating our home here? There's plenty of evidence that humans tend to spend much less effort on maintaining the condition of their tools or environments if they know that replacements are readily available. As an example, consider whether anyone spends as much effort washing and storing disposable plastic cutlery as they do fine china. Is there a danger that this view of space as a replacement for our home planet will accelerate our damage to the environment here on Earth?

There may be individuals today in the space community operating under the impression that we can simply replace the damaged Earth with a fresh new planet. But it seems unlikely that the species as a whole will fall victim to this mindset, given the enormous practical challenges to space settlement. There are certainly untapped resources in space, but nothing that compares to the basic life-supporting resources available on much of the Earth's surface: breathable air, protection from radiation, fertile soil, and easily accessible water. Earth is by far the most habitable place in the solar system, and so far, the most habitable place we know of in the galaxy. Even the planetary scientist who coined the phrase "disposable planet mentality," William K. Hartmann, dismissed the idea that it could gain wide acceptance, as "Earth is the only known place where we can stand naked in the light of a nearby star and enjoy our surroundings."[27] Here, environmental ethics and financial considerations are on the same side: there are clearer benefits and much more predictable costs to mitigating and even reversing the effects of climate change on Earth than to building even one successful, self-sustaining settlement off-planet. Regardless, the choice between settling space and addressing climate change is probably a false dichotomy. Fixing the Earth will likely help us learn how to create habitable environments in space, and vice versa.

From an evolutionary standpoint, ensuring the continuation of our species (specifically, our genetic descendants) *is* the meaning and purpose of life. But as intelligent animals, who can make decisions based on morality rather than biology, we could ask whether perpetuating our genome is worth any cost. Individual humans can and occasionally do make the choice to sacrifice their own lives in order to save the lives of other humans, or even non-human animals. But let's examine that choice, between biology and morality, on a global scale: What if preserving the human species means eradicating or abandoning all other life on Earth? What if it means humankind exists only in a state of misery and deprivation, in an eternally inhospitable and alien environment? This is not to argue that space settlement will definitely result in these worst-case scenarios, but rather to ask whether there is any imaginable case in which allowing or causing humans to become extinct is the more ethical choice.

We're deep into philosophy, now. A common stance among philosophers who ponder this question is that if you assume humans are the only species that have moral values (which is not necessarily an ironclad assumption), then allowing humans to die out would mean that morality itself would cease to exist. Thus, we have a moral imperative, perhaps even our primary imperative, to preserve the human species, thereby preserving morality.

But this is not a universal argument, even among philosophers. Some animal rights activists have argued that nonhuman species have a similar moral value to humans, and therefore the dangers that humans pose to the rest of Earth's species mean that it would be better for us to go extinct, a position philosopher Kelly Smith refers to as "eco-nihilism."[28] Claiming that humans all deserve to die for the damage we have inflicted on our planet is, of course, an extreme opinion. But is there a scenario in which it would be a kindness to allow extinction, rather than a just punishment? For example, philosopher Adam Potthast argues that it may be more ethical to allow a species living a "short, brutish" existence, full of suffering and cruelty, to die out rather than preserve it.[29]

AN ARGUMENT FOR DELAY

WE'RE NOT READY . . . YET

Linda Billings, an expert in social studies of science who has performed science communication research for NASA and written extensively on this topic, believes that our current human spaceflight enterprise is "broken, as in corrupt, like a corrupted file. It's just *broken*. It's bloated, and it's inefficient."[30] She worries that, just as our current climate change crisis disproportionally affects the poor, any attempt to settle space would disproportionately favor the rich, and that those resources might be better spent helping the victims of climate change. But despite these concerns, her position on space settlement is more "not right now" than "never."

"If and when human beings are further evolved, culturally, intellectually, spiritually, in all these important aspects of being human, then perhaps we could proceed," Billings conceded in an interview. "But we're nowhere close right now. We're nowhere close." She estimates that this evolution could take at least a century, and imagines that in such a future, where "we have political and social systems that have improved over time to represent the interests of all people and take care of the needs of all people," we could at that point have an international dialogue about whether to migrate into space.

Billings is not alone in arguing that there is work to be done on Earth before we are ready to move into space. But what does "ready" mean? How do we even begin to set quantifiable criteria on what it means to be an evolved society? Not only do we not all agree on what a good world looks like beyond a nebulous "everyone is happy" but we also don't agree on how we get there.

I put this question to Billings: If we need to wait to go to space until we're better as a species, how do we figure out what "better" means? She agreed that it's a fair question, and that reaching agreement on an answer will be difficult, but that the key is to engage *everyone* on Earth, not just those working in the space industry. "It needs to be, in my opinion, a global consensus. We have no way—at this point and probably not a hundred years from now—of engaging people all over the world who have something to say about this dialogue. And that's what we need."

ASKING THE RIGHT QUESTIONS

A person's stance on whether we should or should not attempt to settle space likely reflects their underlying beliefs and values. If someone is struggling to convert a group of space settlement skeptics, is the problem simply that they don't understand the details of the plan for space, or is there a deeper divide between the prospective space settler's visions of what our shared future should look like and those of their audience?

Linda Billings pointed out that as a trained social scientist, she's had more practice thinking about and expressing her personal values than, for example, a trained astrophysicist like myself. "In the course of my education, I've been forced to identify and describe my worldview and my value systems," she explained.[31] "I've really been pushed to identify the values that I'm applying to whatever writing or research project that I'm conducting. But most people don't do that. And it's no fault of their own, it's just not the way that they're educated. Certainly, in the natural sciences and engineering, there's insufficient attention paid to the need for scientists and engineers to have some better understanding of philosophy of science, the ethics of science."

Does this mean you need to get a social science education in order to work toward space settlement? No, but perhaps the rest of us should spend some time thinking about what our arguments for space settlement indicate about our assumptions and beliefs about our world, our species, and our future. Is progress required for a society's health? Is exploration an intrinsic part of human nature? Do we have an obligation to improve as a society before we spread beyond our home planet?

We may never reach consensus on these questions, despite the rhetoric of people in the space community who imply that they speak for all humanity, or even all life on Earth, when they argue for space settlement. In a practical sense, perhaps it doesn't matter. Because of the shift toward commercial spaceflight, most of the people making plans to actually build space settlements are not trying to build a consensus from anyone except their funders. In this capitalist framework, anti-space-settlement sentiment from an ethical perspective would likely not be enough to stop the space industry. If it's technologically possible and economically feasible, they will eventually send people to live in space.

The current debate around the question "Should we settle space?" does not break down into a simple "Yes" versus "No" divide, but rather "Yes" versus "Not if you're going to do it like *that*!" It may be too late to ask, "*Should* we settle space?" but perhaps that's not the most important question. Instead, we need to ask, "*How* should we settle space?" That's what the following chapters will examine.

2

WHY ARE WE GOING?

The year is 100,000 BCE, and you're standing on a slope overlooking a lush, thriving plain. This is the farthest you have ever been from your home, a fertile valley that has been blessed with plentiful food and an overabundance of babies in the last several years. In your grandfathers' youth, your people's home was surrounded by dry, barren landscapes. But the rains have returned to the land, and as the valley has grown more crowded, the horizon has become more tempting. You and your friends have decided to journey outward, leaving your childhood village behind but carrying the hopes of your families with you, to start a new home among the rich, green hills in the distance.

The year is 2015 CE, and you're sitting on a raft in the Mediterranean Sea. You are six years old. When you got on the boat with your mother, she told you that it wasn't safe to live in your old home anymore, and that you were going to find a new, safe home in Europe. You haven't seen your father in a long time. You wrap your arms around your little brother and close your eyes, praying that you will reach land before the rickety boat tips you all out into the cold water.

The year is 2100 CE, and you're sitting on a rocket. The initial wave of ships will soon leave Earth to establish humanity's first off-world settlement, and you're going with them. But why are you and your fellow travelers here? Are you hoping to seek your fortune in space? To escape your lives on Earth? Or has Earth become inhospitable or even uninhabitable? Maybe you just want to be a part of history, to find a place where you can make a difference?

OUR MOTIVES WILL DETERMINE OUR METHODS

"We choose to go to the Moon," President Kennedy said in a 1962 speech at Rice University, a speech so famous that I didn't have to look it up to quote these words. "We choose to go to the Moon in this decade and do the other things, not because they are easy, but because they are hard."[1] This beloved line is still used today to explain the allure of the unexplored abyss of space, along with Kennedy's quote from explorer George Mallory later in the same speech, who explained his reason for tackling Everest as "Because it is there."

While "because it is hard" was an inspiring and effective motivation for the American public in the 1960s, and remains so for many space enthusiasts today, Kennedy was not shy about another, more urgent reason for sending astronauts to the Moon: competition with the USSR. "No nation which expects to be the leader of other nations can expect to stay behind in the race for space," he told the crowd at Rice. Without mentioning the Soviet Union by name, Kennedy referenced the ongoing ideological struggle between the two superpowers by warning that if America did not reach the Moon with a "banner of freedom and peace," it would be "governed by a hostile flag of conquest."

What motivated humanity's first journey to the Moon: international politics or the spirit of exploration? The answer, of course, is that every individual involved in the Apollo program, from taxpayers to astronauts, had their own combination of reasons for pursuing the Moon. Does it matter what those reasons were? Politics may have fueled the space race, but it was human ingenuity and courage that got us there. In the long run, what difference does it make whether Kennedy was personally motivated by the desire to demonstrate US technological superiority over the USSR rather than the opportunity for scientific discovery and exploration?

Both motivations shaped the Apollo program. Most of the astronauts who walked on the Moon were drawn from the US military, although one was a geologist. The first astronauts to reach the Moon's surface gathered samples and performed scientific experiments, but they also planted an American flag on live television. The clearest evidence that competition, rather than exploration, was the primary motivation for the Apollo program can be found at its close: Humans haven't set foot there since the Cold War ended, even though much of the Moon's surface remains unexplored. When the political motivation dried up, so did the funding and support. Our reasons for space exploration determine the choices that we make there and can ultimately decide the success or failure of the mission.

The same will be true for space settlement. As we saw in the previous chapter, there is no single, universal argument for settling space, even among space settlement advocates. But our reasons for migrating to space will have a lasting impact on the societies we build there, just as the motivations for historical migrations across the Earth have shaped the societies that were formed. Let's revisit some of the motivations from the last chapter and consider the kinds of futures that may result from each one.

SPACE AS (MANIFEST) DESTINY

COLONIALISM AND THE WILD WEST ARE NOT GOOD ROLE MODELS

We saw in the previous chapter that a sense of space as humanity's destiny can be a powerful motivation for space advocates, and that for Americans in particular, this can be mixed up with a veneration of the early American Western frontier. The mythology of the Wild West has even been used to sell space in fiction: *Star Trek* itself was originally pitched as "*Wagon Train* to the stars," referencing a popular and long-running Western that aired in the 1950s and '60s. This "Manifest Destiny in space" is evident in the language we use to talk about space: the "final frontier" of *Star Trek*, the "pioneering" spirit that space settlers will need, and the ubiquitous references to "space colonization."

By now, some readers may have noticed my disinclination to use the word "colony" to describe a permanent human habitat in space. The use of "colony"-related terms in the context of space has become more widely recognized as controversial in recent years due to a spreading awareness of the colonialism entrenched in many of today's discussions of human migration to space. The history of colonization on Earth is inextricably entwined with genocide, oppression, and environmental devastation. And yet, colonialist attitudes still pervade Western culture, keeping many of us oblivious to the fact that colonialism is still causing harm today, in the form of multigenerational inequality, racism, and the ongoing displacement and mistreatment of colonized peoples.

Sociologist Zuleyka Zevallos defined colonialism as being "rooted in historical and political processes. It's really about how various nation-states have been able to enrich themselves through the economic and social control of other countries and other subgroups. And in particular, colonialism is the use of violence and state force as well as ideology that legitimizes taking over the land and resources and cultures of other groups in order to further colonial powers."[2] Let's assume for the sake of this chapter that there is no life in the space environments we hope to migrate to, let alone intelligent life. In the absence of Indigenous people to exploit or harm, does the comparison to colonialism on Earth even apply? How can we be in danger of repeating the mistakes of our past (and present) in space if there are no potential victims there?

When I put this question to Zevallos, she pointed out that colonialist attitudes and structures cause harm beyond the damage done to the colonized people. The process of colonization, she explained, has historically been "one of inherent inequality. The people who finance the colonial efforts are not the people who do the hard work, who will have to build the machines, who will have to build the structures that would facilitate colonization. And certainly, the people who do that labor, that manual labor, will not be the ones who benefit from any space settlements that might be set up."

One of Zevallos's primary observations about space settlement advocates is that they frequently reference colonization without appearing to understand the problematic nature of colonialism. "I think some of that enthusiasm does come from a lack of awareness about the issues that we've faced regarding colonialism in different societies across time," she said. "I think many people who have not been on the receiving end of colonialism don't understand that colonialization is still happening on Earth as we speak."

The language we use to describe our visions of space matters, because it tells us (and each other) what kind of future we want to build. "Colonization" evokes the bold, independent spirit of early American colonists, but it also evokes the enormous inequality and exploitation caused by colonialism. I choose to use words like "settlement" rather than "colonization" to continually remind myself of the dangers of a colonialist approach to space. "Settlement" is also not a perfect, innocent term: for example, it can evoke the conflict over Israeli settlements in Gaza and the West Bank. History is rife with exploitation and conflict over land, and most of our language has absorbed it. Ultimately, the language we use is not as important as the action we take.

There's another aspect of the colonialist space narrative that I hadn't considered until I talked to Darcie Little Badger, an oceanographer and speculative fiction author. Little Badger is Lipan Apache, and I asked for her thoughts on the framing of space settlement as a parallel to the European colonization of North America, and whether it matters that there is no known indigenous life in space. I expected a response along the lines of Zevallos's, pointing out that colonialism can be harmful even in the absence of indigenous life, but Little Badger was instead concerned about the dismissal of indigenous life from the metaphor of the space frontier: "It kind of suggests that the Indigenous peoples in the North American region almost didn't exist. And that's something that I think is problematic, because there is an issue with erasure when it comes to talking about the Native peoples of these lands. There's a tendency to forget that we were here, we had our own civilizations, and we were treated very poorly, dispossessed, murdered, put on reservations. It was genocide. And comparing this land to what is essentially a barren planet doesn't really recognize that."[3]

This is an excellent example of how our work toward space settlement doesn't just affect our descendants in those settlements generations from now but has real effects on Earth today. If we frame space as a continuation of Manifest Destiny but leave Indigenous people out of the narrative because there's no indigenous life to hurt this time, we are perpetuating the effects of colonialism that still plague our civilization today.

The willful ignorance of this framing of space settlement also risks the lives of our descendants in space by neglecting the extent to which historical colonists relied on Indigenous people for survival. In Robert Zubrin's *The Case for Mars*, Zubrin, like many champions of space settlement, emphasizes the importance of building sustainable space habitats where residents can extract everything they need from their environment, rather than depending on shipments from Earth. Also like many space settlement advocates, he invokes the history of colonization on Earth to make his point: "Down through history, it has generally been the case that those explorers and settlers who took the trouble to study, learn, and adopt the survival and travel methods of wilderness natives succeeded where others did not. The foreigner sees wilderness where the native sees home—it is no surprise that indigenous peoples possess the best knowledge of how to recognize and use resources present in the wilderness environment."[4]

Indeed, the history of colonization is rife with examples of colonists benefitting from the local knowledge of Indigenous people, who understood how to sustainably extract resources from their environments far better than the newcomers. American schoolchildren are taught the story of the first Thanksgiving, when members of the Wampanoag tribe shared a feast with the Pilgrim settlers at Plymouth. Half the Pilgrims had died during their first winter in North America, but the Wampanoag taught them how to catch local game and cultivate native crops, and in October 1621 they celebrated the Pilgrims' first harvest together. But the Pilgrims benefitted from more than just the friendship and lessons of the local Indigenous people. In fact, they had chosen to settle at Plymouth in the first place because the land had conveniently been cleared by the former inhabitants, the Patuxet people, before they were wiped out by a series of European diseases brought to the continent by colonists.

Similarly, journals of the first European colonists and explorers in Australia show that the Europeans marveled at the beautiful and fertile landscape, "ornamented with trees, which, although 'dropt in nature's careless haste,' gave the country the appearance of an extensive park."[5] In fact, the park-like appearance of the countryside owed more to centuries of Aboriginal land management and agriculture than to "nature's careless haste." Many of the earliest European explorers of Australia survived their journeys only by raiding the Aboriginal grain stockpiles and fish traps they came across.[6]

This key survival mechanism used by colonists in the past—learning from, trading with, or stealing from the Indigenous people—will not be available to space settlers. By embracing a colonialist vision of space and

romanticizing the pluck and ingenuity of historical colonists while neglecting their dependence on Indigenous knowledge, we are ignoring the
incredibly harsh and dangerous lives that space settlers will experience.
There will be no Thanksgiving feast on Mars, no food caches to plunder,
no locals to teach us how to maintain a balanced relationship with our
environment. We will be on our own in a way that historical colonists
never were.

SPACE AS (FINANCIAL) PROGRESS

THE SHORT ROAD FROM PROFIT-SEEKING TO EXPLOITATION

Why did humans migrate across the Earth historically? There are many
reasons, but one of the earliest was probably the pursuit of resources.
Some groups would migrate seasonally with their animal food source
or move on as the climate changed and food became scarce. But often
people would simply outgrow their homes. As historian Donna Gabaccia
described it: "They out-reproduced their food supply. And their human
groups, which were based on families or wider, more extended notions of
kinship, became too large for the ecological niche in which they had settled. In which case, family groups split, and especially younger members
then moved to the margins, often into new ecosystems and new ecological
niches that required adaptation."[7]

Gaining access to new resources is the primary stated goal of many
of the entrepreneurs in the space industry today. Unlike our prehistoric
migrating ancestors, space entrepreneurs aren't looking to space for
farmland or new hunting grounds to sustain them: they're searching for
something they can sell back to the people of Earth. The (as yet unproven)
economy of private spaceflight is based around accessing resources in
space that we can't get, or can't get as easily, here on Earth.

The pursuit of profit is not a recent motivation for expansion and
exploration. For example, the first known circumnavigation of the globe,
by a Spanish-funded fleet initially led by Portuguese explorer Ferdinand
Magellan, was motivated by the political and commercial desire to find a
western route to the Moluccas. Known to Europeans as the "Spice Islands,"

the Moluccan Islands of modern Indonesia were the sole source of certain expensive spices like nutmeg and cloves in the sixteenth century. Portugal, thanks to the 1494 Treaty of Tordesillas with Spain, held a monopoly on the eastern route from Europe to the Moluccas, and Spain sought its own shipping route. The treaty divided the surface of the Earth in two, granting Portugal the eastern half and Spain the western half, so Spain was also hoping to officially place the Moluccas in the western hemisphere and claim the spice trade for its own. (The rest of the world, including the actual inhabitants of the Moluccas, were not consulted about this treaty.)

Although the Spanish fleet managed to reach the Moluccas and establish contact with the Moluccan people, the western route proved to be more trouble than it was worth: of the roughly 260 men who departed Spain in five ships, only eighteen returned in a single ship three years later, while the rest died or were left behind along the way. The casualty list included Magellan himself, who died in battle while attempting to forcibly convert the people of Mactan to Christianity.[8] Regardless, the first time that humans traveled all the way around our planet like the sphere it is, they did so not for the pure joy of exploration or to find new places to live, but in order to access resources that were scarce and precious in their homeland.

Similarly, the individual companies working to move into space are not doing it purely for the benefit of humankind. Much of the prospective space industry is operating in a capitalist system whose driving force is profit. These companies owe more to their shareholders than they do to you or me, and space is predicted to be a trillion-dollar industry by the 2040s.[9] This profit-based drive has the potential to corrupt our motives for moving farther into space, despite the PR rhetoric of these companies.

One of the darker sides of capitalism, for example, is labor exploitation, which brings us to another reason that humans have moved across the globe historically: because they were forced to. As Gabaccia explained, "Colonization, historically, is not associated with free movement but rather with coerced movement."[10] The primary example of forced migration that comes to mind for most Americans is the transatlantic slave trade. The Trans-Atlantic Slave Trade Database estimates that the Middle Passage of the triangular trade forced roughly 12.5 million Africans across the Atlantic to the Americas (although only 10.7 million survived the journey).[11] Over 300,000 of these Africans ended up in what is now the United States of America, while the rest were taken to Europe, notably Portugal, Great Britain, and Spain. By the end of the eighteenth century, enslaved Africans accounted for about 17 percent of the US population.[12] Within the US, a "Second Middle Passage" in the mid-nineteenth century continued this forced migration, as roughly one million enslaved people were transported across state lines to be sold in areas with higher labor demands, many of them permanently separated from their families.

Today, in this early age of space travel, when the cost of transporting anyone off the surface of the planet is enormously expensive, and more people want to travel to space than will ever be able to afford to in their lifetimes, the idea of space slavers packing Earthlings onto rockets and shipping them into space to mine asteroids for no wages may seem ridiculous. But there are subtler forms of coerced migration. For example, what if those people who yearn to live in space but cannot afford to pay their own way were offered the opportunity to trade their labor and autonomy for passage to Mars? Could this make them vulnerable to exploitation?

In early 2020, SpaceX founder Elon Musk suggested that prospective colonists who could not afford the $200,000 ticket to his proposed future Mars colony could take out loans, to be worked off in one of the many jobs that would be available on Mars.[13] While this proposal aligns with Musk's stated goal to make space settlement as accessible as possible, history provides us with plenty of examples of the risk that such an arrangement can pose to the human rights of the indebted workers.

Consider, for example, the immigrant "redemptioners" in the US in the eighteenth and early nineteenth centuries. Unlike indentured servants, who signed contracts prior to leaving Europe trading their unpaid labor for passage to America, redemptioners arrived in the US already in debt for their journey. If a family member already in the US could afford to pay off this debt, they could present themselves at the dock to "redeem" their loved one. If not, anyone could pay the ship's owner to redeem the traveler, who would then be indebted to their new master and required to work off the loan. The debtor would often be forced to negotiate the terms of their labor contract before they could leave the ship that they had been living on for the past several weeks. Laws in the United Kingdom protected British subjects from these arrangements, leading masters in the colonies to prefer the more exploitable non-British immigrants as laborers. Toward the end of the eighteenth century, nearly half of the German immigrants arriving in Philadelphia were forced to negotiate labor contracts to redeem their transportation costs.[14]

Indentured servitude was abolished in the US along with slavery with the Thirteenth Amendment, but labor contracts are still common in today's world. A worker may sign a contract with a new employer, for example, promising to work for their company for a certain number of years. Unlike indentured servants, of course, workers who sign contracts today can quit whenever they want. Depending on the terms of the contract, they may face penalties such as fines or being blacklisted from working for the company again, but their employer won't hunt them down and force them to return to work. The question for Elon Musk is, what will happen to an indentured Martian worker if they decide to quit? Can they declare bankruptcy and go back to Earth? Given the expense of transporting humans between planets, it seems more likely they would be trapped on Mars. In that case, what's to stop their employers from abusing their unpaid, defenseless workforce in the name of profit, dragging the new society back into the labor exploitation practices of the past?

SPACE AS (SELECTIVE) SURVIVAL

LIFEBOATS ARE NOT THE MOST ETHICAL PLACES

At the moment, humanity lacks the proven ability to move even one human being to Mars, let alone keep them alive indefinitely when they get there. Perhaps one day, far in the future, we will have the technology and resources capable of transplanting the entire population of the Earth to another habitable environment in space. But what if the Earth becomes uninhabitable in the meantime? We'll be left with a lifeboat scenario, in which we will have to decide who gets to escape to space and who will be left behind to die on Earth.

First, there's the question of who gets to board the lifeboats leaving Earth. For a case study of the chaos that might ensue, we have only to look at the famous example of the *Titanic*. It's a well-known part of the tale that there were not enough lifeboats to accommodate all the passengers and crew of the ill-fated ocean liner. This scenario is virtually guaranteed in the event that a catastrophe forces an evacuation of the Earth. Reactions of the *Titanic* passengers and crew ranged from the altruistic, noble, and self-sacrificing to the panicky, violent, and self-centered. Some men gave up their seats willingly for women and children. Others tried to force their way onto the boats and were shot. According to eyewitnesses, one man disguised himself in women's clothing to sneak onto a lifeboat. Another particularly rich passenger paid off crew members to lower his under-filled boat first.[15] It's easy to imagine any of these scenarios playing out in parallel in the event of a planet-wide evacuation of a doomed Earth.

Suppose that by the time disaster strikes and we're loading the spaceships to escape Earth, humans are already living in space. In this scenario, we can imagine the settlements themselves as the lifeboats—small islands of sanctuary with life-sustaining but limited resources—and the refugees in the ships escaping Earth as swimmers in the water, desperate to be let in. If the refugees are allowed in, they will strain the limited capacity of the settlements. But if they are left out in the cold of space, they will surely die.

This metaphor, of nearly full lifeboats surrounded by desperate, drowning swimmers, has been used to make arguments about refugees and immigration on Earth. In 1974, ecologist Garrett Hardin wrote a pair of articles, titled "Lifeboat Ethics: The Case Against Helping the Poor" and "Living on a Lifeboat," arguing that pulling additional victims onto a crowded lifeboat would put the lives of the entire boat at risk, and that therefore wealthy nations (the lifeboats, in his metaphor) should not admit or send aid to the citizens of poorer nations (the swimmers), else they risk suffering their own collapse due to overuse of limited resources.[16] This kind of radical utilitarianism appeals to some, but even today most immigration and asylum policies acknowledge a middle ground in the possibility that sharing some land and resources can benefit everyone, not just those lucky enough to be born in a lifeboat.

If the space settlements reject Hardin's argument and decide to allow the refugees in, they'll encounter the same problems faced by refugees on

Earth today: How do you peacefully integrate into the societies that you're fleeing to? How will our hypothetical space settlers respond to a sudden influx of terrified, desperate Earthlings? Will they welcome the remnants of humanity with open arms? Or will they respond the way many humans today respond to overwhelming numbers of refugees: with suspicion, or even disdain?

Gabaccia agrees that the potential for conflict between refugees and earlier generations of space settlers has plenty of historical precedent: "It's perfectly easy for me to imagine, and you see this in today's world, a real sense of distance and differentiation between the people who are suffering the crisis back there on Earth and the ones sitting pretty, presumably, on their distant planet. It's very easy for me to imagine the folks on the distant planet saying, 'Gosh, those humans on Earth, they just cannot solve their political problems. And why should I solve their problems for them? Why should I send them a ship to bring them to my place? What have they done to deserve that?'"[17]

Gabaccia even suggested a space parallel to the refugee camps that were hastily built across Europe to handle the 2018 refugee crisis: "If I were to project that into the future, what I would see is large numbers of refugees trying to get on whatever transportation is being provided by the wealthy and comfortable interplanetary humans out there, and the remainder dying. Or, alternatively, the creation of an asteroid in between, and 'Well, you guys are going to stay there, because we can't take all of you at once. We'll try to feed you and house you, but you'll have to stay there until we're ready to relocate some of you.'" She pointed out that when conditions in refugee camps become intolerable or even just interminable, refugees will often simply strike out on their own, risking death, arrest, or deportation in an attempt to reach a safe haven. That's dangerous enough now, when the path to safety involves crossing the Mediterranean in an overcrowded boat or trekking across deserts or mountains on foot with no supplies or guide. It would be flatly impossible in space, where "striking out on your own" would require a portable pressurized habitat and a supply of air just to survive the first ten minutes.

If our goal in migrating to space is not the pursuit of destiny or valuable resources but merely survival, we may find ourselves in the position of so many refugees on Earth today: desperate, traumatized, clinging to our children and the hope that there is a safe new home on the other side of the gates, if only the people on the other side will let us in.

WHY WOULD YOU GO?

THE LEGACY OF OUR INTENTIONS

We all search for purpose in our lives, for a reason that we were put on this planet. But our future descendants in space won't have to wonder: their ancestors, the founders of their space settlement, will have put them on their planet for some purpose. How will knowledge of that purpose shape their lives?

A settlement built merely for survival, as a way to avoid putting all of humankind's eggs in one basket, doesn't provide much direction to the settlers' lives. "The meaning of life is to stay alive" is an empty tautology that will likely leave our descendants in space searching for a more enriching purpose. But if they do come to value survival above all else due to the narrative that we pass down to them, what will they lose in the process? A willingness to take risks, to welcome and shelter refugees, to define themselves as more than just vectors for their ancestors' DNA?

What if instead this hypothetical settlement emerges from a commercial enterprise, like a mining operation? The US, as a country, was founded by corporations seeking profit as much as by pilgrims seeking religious freedom, and over two hundred years later, we have yet to rid ourselves of some of the worst aspects of capitalism: inequality, labor exploitation, environmental destruction. How can we expect future space settlers to avoid repeating the same mistakes, if their society starts the same way?

There are other reasons for migrating to space besides the ones discussed in this chapter, and countless worlds that we could imagine resulting from each. The question at the heart of this book—"What kind of world do we want our descendants in space to inhabit?"—requires us to first ask what our intentions are as we start down the path toward leaving Earth. We must be honest with ourselves about those intentions and spend the effort to project them forward in time and space to imagine what kinds of societies they might produce.

Someday, perhaps, your great-grandchildren may be living in one of humanity's first off-Earth settlements. If they could ask you how and why they ended up there, what answer might you give them? Did you bring their grandparents with you to space so your family could be a part of a grand adventure? Did their parents immigrate there to seek their fortune, not considering the implications for their children? Were they sent off-planet in a desperate effort to protect them from a rapidly deteriorating environment on Earth? Now consider how each of these different answers would shape your great-grandchildren's lives, and whether they would thank you or resent you for the decisions you made long before they were born on a planet far away.

3

WHO GETS TO GO?

The year is 1211 CE, and you're sitting on a great canoe somewhere in the Pacific Ocean. You are one of the strongest young men from your village, able to go without food or much water for many days while still keeping up with the other rowers. You were proud to be hand-picked by the chief to accompany him on this long voyage to explore the sea and find new islands. If the land you discover is already occupied, you may decide to live among those people and build a new home and family. Or perhaps you will return to your village with great stories to tell and directions to an uninhabited paradise.

The year is 1787 CE, and you're sitting below decks on a ship just rounding the southern tip of Africa. You've been on this boat for five months and have another whole ocean to cross before you and your fellow transportees reach your new home, the first British penal colony in Australia. Your sentence is only seven years long, but you have no idea if you'll ever see your family again. And you're not sure how exactly you're supposed to help establish this colony, given that your only career experience was as a pickpocket in the streets of London. But hey, nobody asked your opinion about any of this, did they?

The year is 2100 CE, and you're sitting on a rocket. The recruitment posters have gone up everywhere: Be A Part of Our Bright Future! Help Write the Next Chapter of Human History! Mars Needs YOU! You were one of the first people in line to sign up. But what was the next step? Did you need to buy a ticket?

How did you qualify? Did you need experience? Did you have to pass a physical test? And now that you've made it onto the ship, who else is sitting next to you? What do they look like? Do you represent humanity's best and brightest? Most courageous and adventurous? Or just the wealthiest?

———————

MOTIVATION

WHY DOES IT MATTER WHO WE SEND?

Much of the effort and investment of the private spaceflight industry so far has focused on the hardware of space settlement: the rockets that will launch us from our home planet; the long-haul ships that will carry us to our destination; the habitats that will protect us from the harsh environment; and the various technologies that will extract resources, recycle our water, clean our air, and produce our food. But eventually, organizations hoping to build the first human space settlement will need to turn their attention to the wetware component of the mission: the humans who will populate the settlement. In the earliest, riskiest days of a settlement, the capabilities of the human settlers and crew will have just as much impact on the success or failure of the mission as the technology keeping them alive. It will be human ingenuity, creativity, and opposable thumbs that may well save the day when technology breaks down or natural disaster strikes. On the other hand, humans are just as skilled at ruining things as they are at saving the day: we can turn fertile environments into barren wastelands through overharvesting and thriving cities into mass graves through war. We can collapse vast empires with something as simple as an ambiguous line of succession. Moving to the barely inhabitable environment of a space settlement will only intensify the stresses on our hierarchies and social systems, and all the technology in the solar system won't save a mission beset by interpersonal conflict, incompetence, or negligence.

In the longer term, the demographics of the population of a space settlement, especially the first generation, will have an outsized effect on the future of the society. The cultural makeup of the population will be passed down from one generation to the next, and while the settlement will evolve its own unique identity as time passes, the values and traditions of its earliest members will act as the seeds for its future culture. The political ideology of the early settlers will be codified into the foundations of the settlement's government. The settlers' expertise and interests will shape their civilization's economic path. While a space settlement will likely continue to receive immigrants from Earth long after it is established, its founding members will have a lasting impact on the identity of the community. We've seen this happen many times on Earth. Consider the influence of the early colonial Puritan values on today's American culture, or the effects of Australia's history of colonization by convicts and working-class English settlers on so much of their modern national identity.

Given the importance of this first generation of settlers, for both the short-term survival of the settlement and its long-term identity, it's worth asking who will decide which Earthlings to send in our first wave of migration. At the moment, the unspoken rule seems to be that whoever owns the rockets gets to pick the astronauts. Space agencies like NASA have strict selection criteria for determining who they will send into space, even for the handful of space "tourists" who have paid their way. No private spaceflight company has yet released selection criteria for their future pilots and staff: Virgin Galactic currently only employs test pilots, who the company says may one day become "the world's first commercial Spaceship Pilots," and SpaceX capsules have, for the most part, been piloted by NASA astronauts, with the exception of the all-civilian Inspiration4 flight in September 2021, paid for by billionaire Jared Isaacman.[1] Perhaps in the future, once space settlements have been established with stable populations and functioning governments, the settlements themselves will determine who they will allow in, via immigration laws; but at the outset it will be up to the governments, space agencies, and private space companies here on Earth.

It seems likely that the number of volunteers willing to join the first group of settlers on Mars or the Moon will outstrip the settlement's capacity. In 2017, a record-breaking 18,300 people responded to NASA's call for applications to join the astronaut corps (including myself). Only twelve of these applicants were chosen as astronaut candidates (not including myself).[2] Assuming this level of enthusiasm for living in space doesn't diminish in the next few decades, the first serious effort to form a permanent off-Earth settlement will be met with an overwhelming number of volunteers. Imagine being part of the small group of people assigned the daunting task of whittling down a list of hundreds of thousands of hopeful applicants to just one hundred future space settlers. What should we base our decision on?

IDEALISM

WORKING BACKWARD FROM UTOPIA

Let's start at the idealist end of the spectrum. Suppose we have a magic wand that lets us ignore any and all practical considerations. In that case, what's the fairest way to select our space settlers? In this fantasy scenario, I would argue that anyone who wants to travel to space should be allowed and encouraged to do so, but how does this ideal hold up in the real world?

Once we start to view space travel as migration instead of simply exploration, this question starts to look more like a matter of border control than a hiring program for a high-risk job. Is the frontier of space an open border, or should it be regulated? We could take some guidance from the United Nations, which considers freedom of movement to be a human right. Article 13 of the UN's Universal Declaration of Human Rights, written in 1948, states that "everyone has the right to leave any country, including his own, and to return to his country."[3] This wording doesn't concern itself with the direction of travel—given the time it was written, its authors probably assumed that the hypothetical human was leaving the country along the surface of the Earth, not directly up into space—but the Outer Space Treaty of 1967 does specifically address space travel. In Article I, it states that "outer space, including the moon and other celestial bodies, shall be free for exploration and use by all States without discrimination of any kind, on a basis of equality and in accordance with international law, and there shall be free access to all areas of celestial bodies."[4] So, these documents seem to support the ideal that anyone who wants to should be allowed to leave the Earth. In other words, there should be universal access to space.

Of course, there are practical limitations. There's a difference between being *allowed* to travel to space and being *able* to. Having the right to cross a border on Earth doesn't give someone the right to board an airplane or a ship without purchasing a ticket. Backpackers hoping to travel on the cheap won't be able to hike or hitchhike their way into space (despite Douglas Adams's very detailed manuals on the subject[5]). It is—and will be for decades, if not centuries, to come—enormously expensive to travel to space, and it doesn't get any cheaper away from the surface of the Earth. Resources like water, air, calories, and even living space will be unimaginably precious in early space settlements. Trying to cram too many people into such a resource-limited environment will lead directly to suffering and death, no matter what our ideals tell us about who has a right to access that environment. While I wish that we lived in a utopian, *Star Trek*-like future in which every person who wants to go to space is able to, this may never be achievable, and it certainly won't be in the near future.

However, there are still fair selection methods that would let us provide *equal*, if not universal, access to space. For example, we could institute a lottery system. This idea is often suggested in the context of space-as-lifeboat, as discussed in the previous chapter. But let's set the doomsday scenarios aside for the moment and stick to considering a future where Earth remains reasonably habitable and selection for the space settlement program is not required for survival. If it's not a matter of life and death, do we really *need* to ensure equal access to space?

On the one hand, space exploration has always represented some of humanity's most ambitious and idealistic values: peaceful cooperation between nations that are not otherwise allies, curiosity and ingenuity in the face of overwhelming technical challenges, and tenacity and courage in an unforgivingly dangerous environment. In this spirit, we might ask questions like: If space represents a better, brighter future, shouldn't everyone have access to it? Or, at least, an equal chance at reaching it? If space represents the future of humanity, whether better or worse, shouldn't everyone have an equal opportunity to influence and build that future?

On the other hand, the history of space exploration has also reflected the practical limitations and biases of the societies that have participated in it. Not everyone gets to be an astronaut. We don't have a lottery system to select astronaut candidates from the pool of applicants; they're carefully chosen by the space agencies who are building and launching the rockets. The only exception to this, at least so far, is that someone with an extremely large amount of money might be able to convince one of those space agencies to let them hitch a ride. In 1990, the Tokyo Broadcasting System paid Russia at least $12 million to fly one of their reporters to the space station Mir for one week, and between 2001 and 2009, seven private citizens paid roughly $20–40 million each to be flown to space.[6] More recently, American billionaire Jared Isaacman paid SpaceX an undisclosed amount of money for a three-day trip to orbit in the company's Crew Dragon capsule, which seats four. Isaacman chose one of the other three crew members via a raffle, one via an entrepreneurship contest, and one because she was a patient and later employee at St. Jude's Children's Research Hospital, which Isaacman raised funds for via the raffle.[7]

Not only do space agencies have very rigorous and competitive astronaut selection programs but they also place limitations on who can even apply to be an astronaut. For example, applicants must typically be citizens of the country that funds the space agency and meet certain physical and educational qualifications. In their most recent round of astronaut candidate selection in 2020, NASA required applicants to have a minimum of a master's degree in a STEM field, two years of related experience or 1,000 hours of jet pilot experience, and eyesight correctable to 20/20.[8] In the early days of space travel, there were even more restrictions on who could hope to travel to space. Both the USA and the USSR selected their first astronauts from pilots in their respective militaries. As a result, the demographics of the population of people who have been to space are not representative of the Earth's population, either now or in 1961 when Yuri Gagarin became the first human in space. As of early 2021, 569 people had traveled to space, representing 38 nationalities (plus the Soviet Union and West Germany). Only 65, just over 10 percent, have been women.[9]

If we wanted to cling to the concept of fairness, we could insist that every demographic on Earth is represented in space in numbers corresponding to their percentage of the human population. Only 10 percent of space travelers have been women? Let's get that up to 49.59 percent.[10] Out of 100 space settlers, 18 should be from China, 17 from India, and 4 from the United States.[11] But what about the island nation of Tuvalu, which makes up about 0.00015 percent of the Earth's population? Under this system, they won't get a Tuvaluan representative in space until the number of settlers reaches one million. And what if none of the 11,200 Tuvaluans on Earth want to live in space? It wouldn't make any sense to force one of them to move to Mars just to satisfy our demographic goals.

Both universal access to and equal representation in space might very well be impossible. But equal access to space, in which everyone with the desire to live in space has a fair shot at it, is a worthwhile goal.

UTILITARIANISM

BECAUSE ALL THE GOOD INTENTIONS IN THE WORLD WON'T MATTER IF THE SETTLEMENT FAILS AND EVERYONE DIES

It's easy to talk about idealist approaches to space settlement from the comfort of a living room on Earth, but, as is often the case, our best intentions and plans are unlikely to survive first contact with the enemy. (The enemy in this scenario being the cold, heartless, radiation-filled vacuum of space.) Space travel is hard; it's expensive; and for someone who doesn't know what they're doing—or is just unlucky—it's deadly. Fully embracing the concept of equal access to space and running a lottery to select such a small number of candidates could easily result in a group that's not qualified to handle the challenges they will face. We'll have to balance our ideals and values with a utilitarian approach in which we ask what's best for the survival of the settlement. This means selecting candidate settlers based not only on their own desire to go to space but also on what they can contribute to the effort.

Early settlements, before they have had enough time to raise and train the next generation, will need certain occupations and areas of expertise: doctors, engineers, farmers, and so on. It would be poor planning to send an entire fleet of dentists but no plumbers, but it would also be a good idea to bring at least one person trained in dentistry. A crucial step in our selection process must be to determine what range of occupations and skills are necessary for a settlement's survival. Some are obvious: the experts needed for maintaining the health of humans and technology in the settlement. But what about people trained in leadership? In the arts? In childcare or education?

Regardless of which fields we decide to prioritize, I would argue that the most important skills to screen for in a candidate settler are not emergency surgery or spacecraft engine repair, but rather the ability to both learn and teach effectively. Space settlers will need to be able to rapidly adapt to their new environment, absorb information quickly during formal training, and pick up new skills as they live in the settlement. They will also need to be excellent teachers, since they will need to train the next generation to replace them in the settlement. It will be crucial to teach certain skills to as many members of the settlement as possible to guard against the loss of vital knowledge in the event of a fatal accident. Consider, for example, a plague that wipes out all the settlement's doctors early, leaving untrained personnel to treat the rest of the patients. Or an industrial accident that kills the only people qualified to repair the life support system.

For an example of the dangers of losing trained personnel early in a catastrophe, consider the 1967 fire aboard the USS *Forrestal*, a US Navy aircraft carrier operating in the Gulf of Tonkin during the Vietnam War. An electrical fault in one of the aircraft on deck caused a rocket to fire into another plane, igniting the jet fuel and starting a fire that ultimately killed more than 130 sailors on board. The *Forrestal* crew fought to extinguish the fire for hours and eventually managed to save the ship, but the initial explosions of fuel tanks and munitions on deck killed or injured nearly all the specially trained firefighting teams. The remaining crew members did their best, but their lack of training led them to make crucial tactical errors while fighting the fire. For example, one team managed to smother part of the fire by laying down a layer of protective firefighting foam over the burning fuel on the deck, only to have a second crew with a water hose enthusiastically wash the foam away, exposing the fuel to air and allowing it to reignite. Today, all sailors in the US Navy are extensively trained in damage control procedures and the use of firefighting and safety equipment, and a training video containing footage of the *Forrestal* fire is still shown to trainees in the Navy and in civilian fire departments.[12] In a space settlement where an equipment failure could decimate the population in seconds, widespread training in areas like emergency medicine, life support maintenance, and safety procedures could mean the difference between a tragic, but not catastrophic, loss of life and the complete destruction of the settlement.

ACCESSIBILITY AND REPRODUCTIVE FITNESS

BECAUSE WE STILL EXIST IN THE PHYSICAL WORLD

Space agencies, which take a very utilitarian approach to crew selection, consider more than education and experience when reviewing astronaut applications. A major component of the astronaut selection process is the health screening. For example, the European Space Agency required a ten-day visit from its astronaut candidates in 2008 for a medical exam that included everything from physical fitness to dental health.[13] A space agency doesn't want to send astronauts into space who physically can't perform their duties, or who have a higher chance of dying in the harsh conditions of spaceflight. This is both for practical purposes (a dead astronaut can't get the job done) and moral ones.

Is this a model we want to carry on into space settlement? At first glance, it may seem like the most practical and ethical method to screen applicants. Potential settlers with certain preexisting health conditions might have a higher likelihood of death, the argument goes, and depending on the disability, they may not be able to operate the technology required for transportation and survival. For example, spacesuits, also known as extravehicular activity (EVA) suits, are designed for people who have four functional limbs, and who fall within a specific height range. In 2019, NASA had to postpone its highly advertised first all-female spacewalk in part because there were not enough medium-sized, operational EVA suits on board for both of the female crew members.[14] A person with dwarfism, gigantism, or an amputated limb may not be able to operate or even fit into such a suit at all if the equipment available is limited to a small number of generic shapes and sizes.

Up to this point, these arguments have been popular in space programs, which have generally taken many of their procedures and values from the military mindset, and which are responsible for (relatively) short-term missions under limited budget constraints. But one of the key objectives of space settlements is that they be multi-generational: eventually, settlements will not be able to filter the incoming settlers because they will be born into the settlements. We might imagine a science fictional future in which genetic screening and manipulation reach the point where hereditary disease and disability are eradicated, but there will always be the possibility of an injury causing disability later in life.

"Accidents happen," astronomer and disabled rights activist Jesse Shanahan pointed out when I asked her about the challenges of screening potential space settlers for disabilities. "Disability is going to happen in space. It is a fact of human existence. And especially in a hostile, unfamiliar environment, there's even more chance for people to become disabled, both mentally and physically, given the requirements and harshness of the job."[15] This means that no matter how rigorous the medical screenings are initially, eventually we will have disabled people living in space settlements. In fact, Shanahan argues that all humans are disabled in space: it is not an environment we evolved to survive in, and our bodies are not well-suited for low gravity, high radiation, or any of the other harsh conditions of space. Shanahan suggests that this requires a shift in our thinking from medical disqualification to accessibility: "Rather than thinking that we can circumvent the disability issue by just preventing previously disabled people from becoming settlers, instead we need to talk about what kinds of infrastructure we can put in place *because* disability is inevitable." We will all need assistive technologies to help us live in space, so it just makes sense to plan and design our settlements with accessibility in mind. And if we're doing that anyway, there's no reason to exclude people from the initial settlement list due to disability.

Modern disability rights advocates like Shanahan who fight for more accessibility in public spaces often point out that accessibility benefits everyone, not just people with disabilities. The same will be true in space. A person born into a space settlement will need to be able to fit into a spacesuit at any time in their lives, from infancy through old age, in case of pressure loss in the habitat. This means that EVA suits will either need to be manufactured in a variety of sizes or will need to be designed in a manner that can be modified to change size to fit the wearer. Designing EVA suits with adaptable sizes will make space more accessible to settlers whose body shapes differ from the average, and will also benefit any space settler whose body changes size or shape over the course of their life, which is to say, any space settler.

The European Space Agency has recently acknowledged the benefits of developing space technology for a larger variety of body shapes. In 2021, they opened applications for a "parastronaut feasibility project" to astronaut hopefuls with a "lower limb deficiency," leg length difference, or height less than 130 centimeters. "Along the way," states the ESA website about the program, "it will bring innovations and other benefits to the safety and efficiency of future crews."[16]

As another example, suppose we were to design controls for our space habitat that could be used by people with vision impairments: voice controls, for example, or tactile controls with Braille labeling. Many astronauts today experience poorer vision after spaceflight due to the effects of microgravity. If this becomes a long-term problem in a space settlement, accessible controls for people with vision impairments will be crucial. But they will also be useful in an emergency in which people with otherwise adequate vision lose their sight.

In 2014, Italian astronaut Luca Parmitano experienced a leak in his EVA suit that caused water to trickle into his helmet during a spacewalk outside the International Space Station. The water clung to his eyes, making it very difficult to see. Parmitano had to blindly feel his way back to the airlock, grasping clumsily in thick gloves to find the handholds astronauts use to pull themselves along the outside of the station.[17] A similar scenario in a space settlement would be much easier to deal with if the settlement were equipped with controls that do not solely depend on vision. Even in the absence of any vision impairment, voice controls add a measure of convenience to many activities, which is one reason why voice assistants like Siri and Alexa have become so popular.

There is another component of physical health that today's space agencies don't include in their screenings (at least not yet): reproductive fitness. A space settlement will only become truly self-sustaining when it can maintain and increase its population without the need for immigration. Setting aside, for now, the challenges of attempting human reproduction in space, we may need to consider the reproductive abilities of individual applicants and of the population of settlers as a whole. We will have to find a way to balance the needs of the first generation of settlers with the requirements of future generations.

For example, suppose we narrow down our applicant pool to a group of ideal candidates. They have the perfect balance of skill sets, expertise, physical and psychological resilience, or whatever other parameters we have decided are important to us. Now suppose that this group is composed entirely of people who are unwilling or unable to reproduce naturally. This "perfect" crop of first-generation settlers could end up being the only generation! Does this mean that we should reject applicants who are infertile, elderly, or simply uninterested in having children?

My answer is a qualified "no." We should consider the reproductive fitness of the *group* of candidate settlers during the selection process, and perhaps make adjustments if we see that the population of settlers as a whole is unlikely to be able to produce a second generation. But we should avoid including reproductive fitness in our criteria for *individual* candidates. This stance reflects my argument for equal access to space but also, more practically, it acknowledges the alternative reproduction methods available to modern humans. After all, not every member of the settlement will need to be able to reproduce, and the settlement may be able to take advantage of various artificial reproductive technologies we use on Earth today, such as in vitro fertilization or surrogacy.

Besides, restricting our candidate pool to applicants who are both fertile and willing to reproduce biologically will not, by itself, guarantee a stable population. For one thing, we need a large enough population to ensure the genetic diversity of future generations. Small, isolated populations are vulnerable to what's known as the founder effect: certain traits will be overrepresented compared to the parent population, and over time the new population may come to look significantly different from the parent. If these overrepresented traits include recessive genetic disorders, interbreeding within the community will increase the incidence of these disorders in the new population.

The minimum viable human population is estimated to be around 10,000—smaller populations risk inbreeding due to the lack of genetic diversity.[18] This can be decreased by bringing a supply of extra genetic material: frozen embryos, for example, or sperm samples. But we'll also need to plan for accidents and disease decreasing the settlement's population size, and for the possibility that elements of the space environment like increased radiation or low gravity will affect the fertility rate of the settlement. The minimum viable population level of 10,000 also assumes that the population is already reasonably genetically diverse.

If we're already considering examining the genetic makeup of our applicants in order to ensure diversity, it's a short step toward thinking about which genetic traits might be desirable in a space settler. Maybe we should give extra weight to candidates whose descendants would be more physically suited to space. Smaller people, for example, require less rocket fuel to reach space, and they also require fewer calories and less water to survive. Of course, we could also rule out people whose genes produce a disadvantage in space: poor eyesight, for example, or a predisposition toward certain diseases. And then we find ourselves on the slippery slope to eugenics.

Any step in the direction of that slippery slope must be taken with great caution and forethought. For now, I will make the same argument that I did regarding reproductive fitness: we should consider the genetic diversity of our settler population during the selection process, but individual genetic profiling should otherwise be avoided as much as possible.

CAPITALISM

THE RICH WILL INHERIT THE STARS?

I've been pointing to today's space agencies as examples of organizations that select their crew members based on what they can contribute to the mission. But there is an exception to this strict utilitarianism, and as is often the case, that exception involves large quantities of money. A handful of private citizens have circumvented the typical astronaut selection processes by paying a fee of tens of millions of dollars each. Someone feeling generous could argue that this process is still utilitarian, and that one of the things that space tourists contribute to the mission is extra funding. Indeed, many of these fee-paying space travelers reject the term "space tourist" and point out that they completed rigorous preflight training and performed scientific experiments while in space. But the fact remains that they would not have made it to space without the ability to pay. Much of the discussion of passenger selection by today's private spaceflight industry has also focused on financial criteria: selecting passengers based on who can afford to pay for a ticket.

Elon Musk of SpaceX has stated that one of the key steps on his roadmap for space settlement is to bring the cost of spaceflight down to the point where it is accessible to the average, middle-class American. As he said at a talk at the 67th International Astronautical Congress in 2017, "If we can get the cost of moving to Mars to be roughly equivalent to a median house price in the United States, which is around $200,000, then I think the probability of establishing a self-sustaining civilization is very high."[19] His argument, and the argument of many in the private spaceflight industry, is that space settlement must be economically sustainable at every point of its development if it is to eventually become self-sufficient. Given that the only people currently making concrete, near-term plans for off-world resource extraction and habitat construction are private companies rather than publicly funded governmental agencies, it's starting to seem like a capitalist framework for space settlement might be a necessary evil on the path to a multi-planet human civilization. But should that framework include charging would-be settlers for their tickets to space?

In an attempt to answer this question, I'd like to point out some flaws in the SpaceX plan. For one thing, the median house price in the United States is more money than many people on Earth today will see in their lifetimes. The idea that a family will be able to sell their home and other earthly possessions to afford a ticket to space assumes that they have the financial resources of a middle-class family in a developed country. This leaves the rest of the world behind and introduces a lot of the demographic biases discussed earlier in this chapter.

More practically, there are examples from the history of colonization on Earth demonstrating that selecting colonists based simply on their ability to pay can be detrimental, and even disastrous, to the colony itself. One of the more prominent examples is Jamestown, an English settlement founded on an island in the James River in what is now the Commonwealth of Virginia but was at the time the territory of the Paspahegh subtribe of the larger Powhatan chiefdom. While eventually Jamestown became the first permanent English settlement in the Americas, it suffered multiple near collapses early in its history.

The winter of 1609–1610, referred to by some colonists as the "starving time," was a particularly harsh period. At the beginning of this winter, nearly 500 settlers lived in Jamestown. By the following spring, only around 60 had survived the disease and starvation that took nearly all their fellow colonists.[20] These last few stragglers abandoned the colony and set sail for England, although they met a relief ship carrying supplies and additional settlers on their way down the James River and were forced to return.

A number of factors likely contributed to the repeated failures at Jamestown, including natural disasters, conflict with the Powhatan (which later intensified after the colonists killed the Paspahegh chief's wife and children), and bad luck. But the massive loss of life can be partially attributed to the lack of skill and planning among the colonists. Many of the original colonists were upper class, with only a small number of "laborers" accompanying them as servants. These aristocratic gentlemen not only lacked expertise in farming, woodworking, and other vital trades but their privileged upbringings had not prepared them for the physical labor and hardship waiting for them in the colony.[21] The private investors of the Virginia Company, who financed the founding of Jamestown, were impatient to see some return on their investments, and encouraged the colonists to produce goods to be sold back in England or risk being cut off from further support. As a result, early colonists spent much of their time searching for gold or manufacturing lumber rather than farming or constructing homes.

It's not hard to imagine a similar scenario playing out in space. The ingredients are the same: an inhospitable environment, settled by volunteers with more wealth than relevant skills, funded by private investors hoping to exploit new territory and resources for financial gain. It's clear that if we let the financial resources of applicants dominate our candidate selection criteria, we risk repeating the tragedy of Jamestown. But I will go one step further and argue that the ability to pay should not be considered during selection at all. Should an applicant who is selected to move to a space settlement be allowed to liquidate everything they own and contribute those funds to the company or agency helping them get there? Sure, why not. But we should not judge our candidates based on their ability or willingness to do so. Limiting our settlers to only those who can afford the trip violates both the ideal of equal access to space and the practical approach of prioritizing the most qualified candidates.

If one day I'm lucky enough to board the first spacecraft taking settlers to Mars, I don't want to see the passenger compartment filled with only the rich and powerful. I want to look around at the people I will be spending the rest of my life with and see fellow passengers from all corners of the globe, people who will work with me to keep the settlement alive, and to teach our children the skills and values they will need to build a new society that I can be proud of. When I imagine myself making that first journey to humanity's new home, it is in the company of Earthlings from six continents, doctors and engineers, plumbers and farmers, leaders and teachers, children and parents, and, if I'm very, very lucky, my dog.

II

HOW WILL WE LIVE WITH THE LAND?

4

WHO OWNS SPACE?

The year is 1450 CE, and you are at home in your village at the edge of a field in central England. Your family has farmed this patch of land for centuries, and your children will farm it after you are gone, but the land itself does not belong to you. You are merely a tenant on the lands of a lord who allows you to farm the land in exchange for a share of your harvest. The land does not belong to the lord, either; it was granted to him by the king in exchange for his military service. And all the land of the kingdom was bestowed upon the king by God himself. Nevertheless, today you will rise from your small bed and work this tiny piece of God's land with your own hands.

The year is 1865 CE, and you and your family are finally free from the slavery you have lived under your entire lives. But now you need to recover and build a life for yourselves. You're a strong worker, and you know how to farm this fertile country; you've done it all your life. The government has taken land from former slave owners and parceled it out to freed families like yours. You finally have a place that is yours, and you all work hard to make it a home. But now a rumor is spreading that the new president wants to take that land back, land you've been working since you were an enslaved child, land that you finally owned in your freedom. How will you start over again? Where will you go?

The year is 2100 CE, and you've finally reached your new home among the stars. You take your first step off the transport ship from Earth, breathing in the air of the habitat. You're ready to set down your luggage and relax after your long journey. But where can you do that? Will you be assigned a room in a shared habitat, or were you expected to purchase private land for yourself in advance? If you're planning on making your fortune here by mining or farming, how can you get the rights to work that territory? Who owns the land beneath your feet? Is it even legal to own land in space?

––––––––––

MOTIVATION

WHY DOES PROPERTY MATTER?

Just as on Earth, different locations in space will grant various advantages to those who have access to those places. Currently, many investors and entrepreneurs are eyeing space for the valuable physical resources that they hope to extract and then sell at a profit, like water from the Moon or precious metals from asteroids. But space also harbors other, less tangible resources. For space settlement advocates concerned with overpopulation on Earth, outer space means literal space: room to grow and expand our civilization. For others, space offers advantageous or strategic positioning. Future settlers will be able to build communities far from terrestrial enemies or oppressive governments. Certain orbital configurations, the intangible opposite of physical territory on the surface of a planet or moon, represent locations where a space station can maintain a comfortable stationary position, or a way for an interplanetary spacecraft to reach its destination with a minimal fuel requirement. Even in low Earth orbit today, different orbits provide better views of areas of scientific interest on the Earth, or a peek into other nations' backyards. As we continue to move outward beyond our planet, developing new technology to improve humanity's ability to access this extraterrestrial territory, how will we decide who will benefit from its advantages?

If I stuck a pin in a map of any continent on Earth, odds are good that at least one person, corporation, or government claims that land as their own. Even Antarctica, the only continent without an Indigenous population, is almost completely blanketed by a quilt of overlapping claims by various nations. But property in space, including land on the surface of bodies like the Moon or Mars, is not generally considered to be "owned" by any individuals or groups on Earth.

Until recently, there hasn't been much point to claiming territory in space; after all, what good would it do anyone to claim ownership of property that they can't use or even visit? For nearly all of human history, space was accessible only through remote observation, and was thus equally available to anyone who had the ability to observe it. There has certainly been conflict over who owns specific observatories and the pieces of land they occupy here on Earth, but in general, one stargazer's observation of the night sky doesn't detract from their neighbor's enjoyment of the same view. But as humans continue their expansion into space in the hopes of establishing permanent settlements and accessing valuable resources, conflicts will arise over who has the right to extract those resources, build structures, access certain areas, and exclude unwelcome visitors. Humanity has struggled through countless similar conflicts over land on Earth throughout our existence, and this history has taught us that having a consistent property rights system, one that everyone agrees on, is crucial for a number of reasons, all of which will apply in space.

THE BENEFITS OF PROPERTY RIGHTS

MORE THAN JUST A PLACE TO BUILD YOUR HOUSE

Economists argue that one of the primary benefits of property rights is, naturally, related to economics. Specifically, private property rights are considered a key requirement for capitalism, the system under which most spacefaring companies are currently developing their space programs and industries. Property in space, along with the rockets, mining equipment, and other technologies required to extract resources, is a means of production. Applying labor to property produces valuable goods—water and precious metals, in the case of space mining, for example—that can then be sold to produce wealth. But without the investment of labor, the resources just sit there, producing nothing valuable (in the economic sense). In order to incentivize people to invest the necessary effort to improve the land and extract its resources, the argument goes, they need some kind of guarantee that they'll be the ones to benefit from those resources. Private property ownership is that guarantee. It doesn't just give a landowner access to the property; it also gives them a sense of security that they'll still have access to the property next week or next year. This lets them plan for the future.

"People will invest in a resource only if they are secure in their ownership or future use of that resource," property lawyer Yuliya Panfil explained during a 2019 panel on the legal landscape of space. "A lot of my work in the past has been with cocoa farmers, for example. A farmer won't plant a tree on land that they have insecure rights over, because in the ten years that it might take the tree to grow, that farmer might be kicked off their land, and then their investment will all be for naught."[1] Panfil also pointed to another example that is probably more familiar to non-farmers: "The types of upgrades that homeowners make to their homes are quite different than those that renters will make, because a homeowner knows that they will be staying in that home long-term."

The complexity of our economic system means that a landowner doesn't even need to plant crops or mine the ground to make a profit from their land. "A lot of economists point to property rights as one of the cornerstones of wealth creation," explained Laura Montgomery, a lawyer who specializes in space policy and law. "Because you can leverage those rights, you can take out a mortgage, you can get a loan off of it, you use it as collateral. And then you can make it grow from there."[2]

This economic perspective has been explicitly cited by companies advocating for private property rights in space. "Without property rights, any plan to engage the private sector in long-term beyond LEO [low Earth orbit] activities will ultimately fail," states a report commissioned by NASA in 2013.[3] The report, produced by space station module manufacturer Bigelow Aerospace, argued that in order to develop a stable space economy, companies must have confidence that "they will be able to (1) enjoy the fruits of their labor relative to activities conducted on the Moon or other celestial bodies, and (2) own the property that they have surveyed, developed, and are realistically able to utilize."

Many economists and environmental conservationists also see private property rights as an important incentive for conserving resources. The argument is that if a limited resource, like a common grazing area or fishing stock, is shared without restriction, then the individual users will be incentivized to take as much as they can without consideration for the collective harm they are doing, leading to the destruction of the resource. This is the "tragedy of the commons," a term coined by ecologist Garrett Hardin: a behavioral phenomenon blamed for climate change and other forms of environmental destruction.[4] In space, this tragedy might take the form of Martian settlements collectively depleting the frozen water in Mars's polar ice caps as they mine them for drinking water. Closer to home, the increasing number of decommissioned satellites in low Earth orbit is already polluting the orbital environment with dangerous debris.

Many economists, conservationists, and public policy experts, including Hardin, have argued that the tragedy of the commons is an inescapable consequence of human behavior that can only be overcome by one of two methods: regulation of the commons by an external authority, or privatization of the commons. The argument for the latter approach is that if the resources are distributed among consumers, who are each given full ownership of their portion, then each individual owner is incentivized to preserve their share of the resource.

This perspective on human behavior was challenged in 1990 by economist Elinor Ostrom, whose Nobel Prize–winning work pointed out that the conclusions drawn from the tragedy of the commons model are not borne up by observations of the way humans actually act in a commons. Ostrom studied examples of a type of commons that she called a "community-pool resource": a resource that is finite and can therefore be used up if too many users try to extract value from it too quickly, but also one that is large enough to make it difficult to keep out "free-riders," users who extract value without putting in their fair share of labor to maintain the resource. Ostrom discovered that while some community-pool resources are depleted due to self-interest, others are well maintained without the need for either a strong external authority or full privatization.[5] This indicates that there may be alternative approaches, besides private property, to protecting the space environment from overuse.

Beyond the practical economic considerations, there are also sociological components of property. Arguments over property on Earth can get deeply heated, whether it's two neighbors sparring over their fence line or two countries going to war over a contested border. Where do these strong feelings come from? Through her work, lawyer Yuliya Panfil has observed the powerful emotional and cultural connections that humans have to their land. "For many people around the world," she said, "land is not just the place where you put your house. It's your unit of belonging in a society. It's the unit through which you measure your culture and your society's history."[6]

Debbie Becher, a sociologist at Barnard College at Columbia University, pointed out another source of strong emotion regarding property conflicts: a sense of fairness and justice. Property rights systems, she noted, "are social agreements about who gets what . . . Agreements over rights, over all kinds of things that we might claim to resources. And so I think people have strong feelings about who deserves what, even if they're not personally involved."[7] This can help explain why the debate over property rights in space can bring up strong emotions even in people who will never live in space themselves. If space is going to be shared among our descendants, we want to see it shared fairly.

What does a fair distribution of land look like? Legal considerations aside (for the moment), there are plenty of options for how to divide up property in space. Different nations could claim different territories, as we see in Antarctica. We could conduct a lottery or auction off parcels of land to the highest bidder. We could simply grant property on a first-come-first-served basis, where the first person to reach the surface of another world can claim as much land as they want. Without an agreed-upon property distribution system, this last option will likely occur by default. After all, if the first settlers to reach Mars insist that they own the land they build their settlement on, no one else will be on Mars to stop them.

"Do we really, as a world, want to reward whoever gets there first?" Becher asked. "I think some people do, because it's an incentive. Clearly that's part of the U.S.'s interest, is creating an incentive for their companies to get there."[8] How do we want to distribute space property? Is a distribution of private property rights even the right approach? As a world, we should make these decisions now, while all the stakeholders are still here on Earth.

WHAT IS PROPERTY?

OUR RELATIONSHIP WITH "STUFF"

For most people, the word "property" just refers to "stuff": land, yes, but also the structures on that land, as well as physical possessions and even intangibles like ideas or radio frequencies. If you can own it, or claim some kind of right to it, it's property. But sociologists and anthropologists have observed that in human societies, property rights systems aren't really about "stuff"; they're about people. More specifically, a property right isn't a relationship between a person and an object, it's an agreement between people *about* the object. And these agreements aren't limited to written deeds and titles, or other property laws in a society. Becher explained that a sociologist like herself "would see property as a kind of social agreement. Property rights certainly exist in writing, in law, and that's part of what they are. But they're really only real to the extent that we respect them. [So, property rights are, in a sense,] claims that people make that get respected by others. So they're always social."[9]

From a legal perspective, property is considered a "bundle of rights," with various legal statuses representing different combinations of rights. For example, if I own a house and my neighbor rents the house next door, we both have the right to live on our respective properties. But as an owner, I can sell my house, while as a renter, my neighbor can't sell theirs. The modern Western property system tends to emphasize ownership as a way to hold the entire bundle of rights to a property, but in practice, even privately owned property can have multiple overlapping rights held by different people. In the previous example, even though I own my house, third parties may also have rights to my land—airplanes are allowed to fly over the house, and public works might dig up my yard to maintain utility lines. My government may even have the right to take my land for public use, under what the US calls eminent domain.

Space settlement advocates tend to view space as a commons, with ownership shared by every human on Earth. On the other hand, some space lawyers (lawyers who specialize in law and policy regarding space) like Henry Hertzfeld and his colleagues argue against this interpretation of current space law: "All too often, commentators and pundits remark that outer space itself belongs to everyone. It is in fact just the opposite. Space itself belongs to no one and the right to access, explore, and use space is granted to everyone."[10] Even if everyone agreed on a single framework for space property rights today, this bundle of rights—access, exploration, and use—may very well shift as laws evolve over time.

Clearly, when we talk about property rights in space, it's more complicated than just working out who owns which piece of extraterrestrial land. Perhaps we'll end up with a system where enclosed habitats, along with the air and water they use to keep their inhabitants alive, are considered common property shared by all residents, while mining sites are private property that can be bought and sold. Alternatively, maybe a settlement will have a central government that grants or sells mining rights for certain regions without selling full ownership of the land.

HISTORICAL COLONIAL LAND DISTRIBUTION

LESSONS FROM OUR TROUBLED PAST

As we think about how to distribute property rights in space in the future, we could look to the past for inspiration and ask how land was distributed in historical colonies. The first step, of course, was for the colonizers to assert that the land was "unowned," which was often false. Colonizers tended to acknowledge that Indigenous people were present on the land that they wanted to colonize, but usually argued that the Indigenous people weren't using it properly, and therefore didn't own it. As justification for this argument, advocates for English colonization of North America often referenced the work of philosopher John Locke, who considered private property to exist only where human labor has been applied to the natural world. By "labor," Locke and his contemporaries specifically meant agricultural cultivation and enclosure.[11] The Indigenous people encountered by English colonists did not fence in their farmland. Thus, in the view of the colonizers, the land was "*terra nullius*," a Latin term meaning land that is empty or unoccupied. Obviously, Indigenous people *were* using the land, they just weren't using it in the way that the colonizing societies considered "civilized."

Centuries later, similar justification was used for the Dawes Act, which subdivided Indigenous tribal land into allotments distributed to Indigenous families and individuals. Advocates of the legislation argued that forcing Indigenous people to implement private property rights would allow them to use their land more productively and to better assimilate into US culture, ignoring the fact that the Indigenous people already had their own perfectly functional property law that simply differed from the US system.[12]

Why does this matter in space, where the land is *actually* not being used by any Indigenous humans (or aliens, as far as we can tell)? Even if space truly is "empty land," the historical appropriation of land by colonizers, based on their belief that Indigenous people were not using their land "correctly," provides a crucial lesson about property rights: different cultures have different approaches to their relationship with land.

Once land was declared (by the colonizers) to be unoccupied and was claimed as the territory of the colonizing nation, it had to be distributed among the colonists somehow. In North America, a large swath of the east coast was claimed by Britain, and then granted to two large corporations by the British king: the Virginia Company of London and the Virginia Company of Plymouth. These corporations, known collectively as the Virginia Company, were then faced with the task of making a profit off the land. Disappointed by the lack of gold, which Spanish colonizers were currently seizing from the Indigenous peoples of South America, the Virginia Company turned to tobacco, a valuable crop that required large amounts of both land (which the Virginia Company had in abundance) and labor (which they did not).

"To try to induce people to come, they started offering shares of land to people who brought labor with them," colonial historian Margaret Newell said. "So you got a share if you came yourself, but you got the same amount of share, and *you* got the share, if you brought laborers with you, and that included indentured servants, enslaved Africans, or your own family members. These were called 'headrights.' So this was actually a company decision. But it became very customary."[13] Corporations hoping to earn a profit by extracting valuable resources from space property may face a similar labor challenge. At the start of space mining and settlement, there will be significantly more land than labor. Systems like the headrights system of the Virginia Company, applied to space, might easily incentivize labor exploitation like indentured servitude and slavery.

This property-granting system also led to unequal distributions of land. A landowner who could afford to pay to import extra labor could gain more land, increasing their own wealth. The Virginia Company failed in 1624 and Virginia became a royal colony instead, but its right to self-govern was maintained, and the consolidation of land continued. "The local governments that emerge out of these corporations—and they had a lot of leeway, and a lot of self-government—they pretty much tended to give land to each other," Newell explained. "That was essentially one of the functions of local government, was to give land and get land. So the leadership cohorts of all of these colonial ventures very quickly became owners of lots of large estates. They got land as rewards, gave land to each other."[14]

This uneven distribution of land was perpetuated over subsequent generations by inheritance and the continued distribution of land to people who were already landowners. A family descended from a poor immigrant could also gain land over time, if their farm was profitable enough that they could purchase more land and afford the labor to maintain it. But families that were granted land early in the colony's history had a big head start. Many people living in North America didn't even have the legal right to own land at the beginning of colonization. Indigenous people certainly weren't allowed to participate in this system, and neither were the enslaved Africans who were brought to the continent so their captors could gain property through headrights. Even European women living in colonial America could not own land under British law. Given the potential for wealth creation that private property provides in a capitalist system, these inequalities compounded over time, even after women and non-white Americans were eventually given equal rights to own property in the US. For example, the lack of access to land ownership for enslaved and, later, freed Black Americans contributed to the racial wealth gap that exists between Black and white Americans today, although it is difficult to determine the degree of effect this had when compared to the racially discriminatory policies and laws that have persisted in one form or another in the US since the abolition of slavery.[15] Even globally, countries with a history of high levels of land ownership inequality correlate with relatively high levels of income inequality today.[16]

It's clear that the way that land is distributed early in a colony can have generationally lasting effects on inequality. But if we don't get it right the first time, at the start of space settlement, surely we can simply adjust the distribution of property once we recognize our error, right? History states otherwise. In fact, history shows that land reform and redistribution is extremely difficult and particularly prone to corruption when financial motives are involved. After the US Civil War, Union General William Sherman ordered 400,000 acres of confiscated Confederate land set aside for formerly enslaved Black refugees. Each family was to be granted up to 40 acres, allowing them to live independently and begin to rebuild after generations of enslavement.[17] Less than a year later, President Andrew Johnson overturned the order, returning the land to its former Confederate owners. The freed refugees were left without land once again, and many were forced to return to working on the plantations where they had been enslaved, this time as sharecroppers. Rather than becoming an example of successful and restorative land redistribution, the phrase "40 acres and a mule" (the mule referring to leftover Union army mules given to the refugee families) has come to represent the broken promise of the US government.[18] The frequent failure or corruption of land reform throughout history teaches us that we need to think carefully about how property rights will be distributed in space now, in these early days of mining and speculation, and how that might affect residents of space settlements in the more distant future.

CURRENT LAW

AN INTRODUCTION TO THE OUTER SPACE TREATY

As we study the past to figure out how property in space will be or should be distributed in the future, we also need to ask what current laws say about space property rights. Most conversations about space property law—or any space law topic—start with the Outer Space Treaty of 1967. Formally titled the "Treaty on Principles Governing the Activities of States in the Exploration and Use of Outer Space, Including the Moon and Other Celestial Bodies," the Outer Space Treaty was written to provide an initial legal framework for how nations can behave in space. The treaty covers liability for crashed spacecraft, the use of nuclear weapons in space, and potential harmful contamination of moons and other planets.

One of the main purposes of the treaty was to prevent another wave of colonization and imperialism, this time in space, as nations developed the technologies to send humans off Earth. Article II addresses this directly, stating that "outer space, including the moon and other celestial bodies, is not subject to national appropriation by claim of sovereignty, by means of use or occupation, or by any other means."[19] Thus, a nation like the UK or Spain cannot claim territory on the Moon or Mars the way they did in the Americas during the colonial period.

But recall the example of the Virginia Company: Initially, the colony of Virginia was founded by a corporation with a royal charter, not by the British government directly. Similarly, it is private corporations who are now expressing interest in traveling to space to "use or occupy" property through mining or settlement. When Moon Express was granted approval in 2017 to land on the Moon, the company's co-founder boasted that "only three super powers have ever landed on any planet—we will become the fourth super power."[20] While Moon Express failed to accomplish its planned landing in 2018, it seems likely that in today's era, when corporations can be richer than countries, settlement in space may be driven by privately funded organizations rather than by national governments.[21] Article II of the Outer Space Treaty only forbids national appropriation of territory; what about private appropriation? Can private companies own property in space, or at least own the resources they hope to extract?

The only mentions of "non-governmental entities" in the Outer Space Treaty exist in Article VI, which requires that terrestrial governments supervise the activities of such entities in space but neglects to discuss their legal ability to claim property. The treaty was written at a time when space exploration was performed solely by government agencies, and private space companies were virtually non-existent. Today, questions about what private companies can and cannot do in space have become more urgent, but the ambiguity in the Outer Space Treaty remains. In a common-law system like the one used by the US and the UK, questions like these are answered through court decisions. Here in the early days of the private space industry, those court cases haven't happened yet.

"Legally, we are at a crossroads in interpreting the 1967 Outer Space Treaty," space lawyer Laura Montgomery said, "because what we have are what we lawyers call 'questions of first impression.' We don't have a lot of court decisions—well, we have no court decisions—interpreting anything in the treaty. And so while we may have positions that the government has articulated, or hopes like space enthusiasts have, we don't have any definitive answers from a court."[22]

In the meantime, subsequent treaties and laws have been written to attempt to clarify the rules for private entities acting in space. For example, in 1979, a United Nations committee developed "The Agreement Governing the Activities of States on the Moon and Other Celestial Bodies" as a follow-up to the Outer Space Treaty of 1967. The Moon Agreement, as it is known, specifically prohibited private ownership of land on the Moon in Article XI, which states that lunar territory cannot be "property of any State, international intergovernmental or non-governmental organization, national organization or non-governmental entity of or any natural person."[23] This would have prevented any private property ownership in space and negated a good chunk of this chapter, but the Moon Agreement was ratified by only eighteen countries, none of which is spacefaring (compared with the 110 nations that have ratified the Outer Space Treaty), and is generally considered to be a failed treaty. The question of the legal status of private property in space remains murky and untested.

Someday soon, if things go according to plan, private companies will have the ability to leave Earth and occupy the land they want to exploit in space, potentially allowing them to claim ownership through squatter's rights. But the more immediate concern for space law right now is to determine whether resource extraction is allowed under the Outer Space Treaty, regardless of whether the land itself can be owned. This has become especially urgent as space mining companies have begun trying to attract investors. To claim that they'll make a profit, they need confidence that they'll be allowed to own, and then sell, resources that they extract from space. Bob Richards, CEO of Moon Express, illustrated this focus on resource ownership rather than land ownership by comparing space mining to fishing in international waters on Earth: "They don't own the water and they don't own the fish, but they have the right to put the nets into the water and bring the fish onto the decks, and once the fish are there, they own the fish."[24] Richards proposed that space mining should work the same way.

In 2015, to address the question of resource ownership and encourage growth in the private space industry, the US government passed the Commercial Space Launch Competitiveness Act. To avoid running afoul of the Outer Space Treaty, the law carefully states that it does not represent a claim of "sovereignty, or sovereign or exclusive rights of jurisdiction over, or the ownership of, any celestial body," but it does specifically assert that a US citizen engaged in resource extraction in space "shall be entitled to any asteroid resource or space resource obtained, including to possess, own, transport, use, and sell it according to applicable law."[25] This law is not without controversy, as some argue that it represents a claim of sovereignty by the US over space resources.[26] However, Luxembourg passed a similar law in 2016, and other spacefaring nations, including Russia and China, are working to pass their own laws protecting claims on space resources. As the private space industry grows, this debate about private ownership of space territory and resources will continue to unfold.

ALTERNATIVE PROPERTY SYSTEMS

OTHER WAYS OF KNOWING THE LAND

The system of private property ownership that dominates Western societies on Earth today is not the only possible model for property. In fact, it hasn't even been around that long. For much of human history, land has tended to be attached to families or other kinship groups, passed down from generation to generation. Selling or otherwise "alienating" the family land to non-members was considered unusual or even forbidden. The Old Testament of the Bible contains several references to this model, in the form of God's orders to his people not to sell the inheritance he has given them.[27] Societies with this type of property system consider land to be owned by future descendants as much as by the family's ancestors and the current generation; therefore, selling the land deprives those unborn descendants of their birthright. English society, followed later by the rest of Europe, gradually evolved away from this model of inalienability between the thirteenth and sixteenth centuries, toward a system where landowners were legally and culturally free to sell their land.[28] This is the model of property rights that they carried with them during their colonization of North America and other parts of the world. The result was the modern Western conception of land as a kind of property that can be privately owned by an individual and exchanged for money or other valuables with other, unrelated individuals.

To consider how we could use and share extraterrestrial property, we should look beyond the current Western, capitalist system in which much of the space industry is developing, to seek examples of what might work better (or possibly worse, of course) in space. We could, for example, replace individual private property rights with communal or state ownership, as suggested by economic ideologies like socialism and communism that developed in response to capitalism. In the words of Karl Marx himself, "The theory of the Communists may be summed up in the single sentence: Abolition of private property."[29] Most countries today use mixed economic systems that lie somewhere on the line stretching from pure capitalism to pure socialism, but we're not even restricted to this spectrum for our space property system. For examples of alternatives, consider the Indigenous property systems that the Europeans violently displaced during their colonization of the Americas.

"[Historians] are more and more understanding the Indians did have firm ideas about territoriality," Newell said. "They fought to defend. They were interested in resources. They fought to defend access to resources. They were particularly interested in transportation bottlenecks: portages, or places where rivers connected, places that actually drew a lot of Indians in for trade . . . They're aware of territory and resources, and they care about them. They care about transportation networks and controlling them."[30]

However, while the Indigenous people of North America had complex relationships with land and with other Indigenous nations about how to share resources, they did not, for the most part, practice the concept of permanent, exclusionary land use, in which a buyer of land could forever after prevent access by the previous owners. After all, alienation was a fairly recent invention in Europe at the time of North American colonization, and most Indigenous societies were not exposed to the idea until contact. English colonizers ultimately benefitted from this confusion and miscommunication, according to Newell:

[Indigenous leaders] would sign agreements conveying what they thought were use rights to the Europeans. So . . . they'd be signing a deed and what they thought was, "You get to do this, too." And the Europeans would be thinking, "I have, now, exclusive access and if you come on this now, you're trespassing." So this notion of exclusivity and of land as something that can be conveyed . . . that you could alienate land, that it was alienable. In other words, that once you sold it, you could never go there again[, you were then alienated] from it . . . It's sort of a new idea for everybody . . . It's not entirely the way things are being practiced in Europe, either.[31]

Indigenous writer and activist Julian Brave NoiseCat agreed that while land use and inheritance systems among Indigenous societies are extremely diverse, "Generally speaking, it's safe to say that a more property-dominant system of land management was not common to most Indigenous peoples in North America. There was a notion of territory, there was a notion of family and hereditary rights to place. And there was some notion that different pieces of land could belong more to one people or one family than to other people. But the Lockean conception of property and property rights is definitely not something that existed here. There was much more of a notion of commons."[32]

Indigenous societies developed a variety of diverse property rights systems, adapted to their unique environments and resources. They understood and incorporated concepts of private property, but rather than strict, alienable "ownership," these systems tended to emphasize use rights: who could hunt or fish in certain territories, or how the crops from a piece of farmland would be shared.[33] As an example, NoiseCat described a 1701 CE treaty between two Indigenous societies: "The Haudenosaunee and the Ojibwe people had an agreement that this would be a common hunting ground, that the area that is now Toronto would be a common hunting ground. And similar agreements existed between First Nations all over the place, where a certain area would be understood to be a commons, in a way." The principle on which this treaty was built was called "Dish with One Spoon," referring to the shared nature of the hunting ground, and the importance of each party to the treaty taking only what they needed from the communal "dish."

Indigenous people, then, were aware of the risk posed by the tragedy of the commons, and they worked to counteract it through law and custom. This is not to say that Indigenous people never overused their resources. But over the generations, many learned to manage those resources sustainably. In his book on pre-contact societies in the Americas, *1491: New Revelations of the Americas Before Columbus*, author Charles C. Mann describes the carefully managed landscape of southeastern North America: "the surgery was almost without scars; the new landscape functioned smoothly, with few of the overreaches that plagued English land management. Few of the overreaches, but not none."[34]

"The soft thesis that Charles Mann advances in his book," NoiseCat noted, "that I've seen people make elsewhere, is that those experiences of resource depletion basically led to reforms of Indigenous growth and production, if you will." Europeans, on the other hand, tended to respond to their depletion of their own resources by continuing to expand outward, seeking new resources through colonization and conquest.

NoiseCat also pointed out another difference between Indigenous and European property systems. "Some First Nations . . . say that their people also had treaties and agreements with different animal nations about how many of the elk or the deer or the bison or the whatever they would take. And I guess that understanding of humanity, as more animal-adjacent than distinct, is obviously quite different from most Western conceptions." Indeed, while many Western legal systems today recognize corporations as people, the personhood of animals, even the most intelligent species, is controversial. Negotiating with an animal nation is far outside the Western worldview.

However, modern governments are beginning to explore the idea of extending legal "personhood" to non-human entities. In 2017, New Zealand's parliament passed a law recognizing the Whanganui River as a legally protected, living being. The law was the result of over a century of advocacy by the Māori people who have lived on the river's banks for centuries and consider it sacred, and this change in the Whanganui's legal identity brings New Zealand law into alignment with the Māori understanding of the river's status, as well as providing it with legal protections. Under the law, two guardians must be appointed to represent Whanganui in court, one Māori and one from the New Zealand government. These representatives will help defend the rights of the river itself as ongoing arguments about human use rights—by hydroelectric companies, farmers, and local residents—continue to rage.[35] This example shows how Indigenous conceptions of land can be incorporated into non-Indigenous legal systems. Such laws may provide us with a template for incorporating multiple perspectives on land use into a single, consistent property rights system in space.

WHAT WILL LAND MEAN TO US IN SPACE?

HOME SWEET (HABITAT) DOME

Unlike many of the topics covered in this book, property law in space is a well-studied issue. International laws defining property rights in space already exist, including the Outer Space Treaty, but the recent growth of private companies hoping to make a profit from space mining has served as a reminder that these laws are not comprehensive, particularly when it comes to the rights of individuals or corporations rather than nations. Sociologist Debbie Becher argues for determining a consistent property rights system in space sooner rather than later. "The lessons from colonial times as well as from today [are] that when claims are ambiguous, that people physically go get the stuff, and then they later say, 'It's ours.' And often, that's respected. So, in some ways, it might be that the earlier that we establish what the law of property is, or what the rule is, the less likely people are to have what we might think of as a 'Wild West' attitude towards it."[36]

Suppose we decide not to simply copy the Western private property system that dominates our terrestrial legal system into space? Maybe this moment is our opportunity to consider alternatives, to explore the perspectives of human societies, both now and in the past, who practiced different ways of living with each other and the land. Space archeologist Alice Gorman, for example, has suggested following the example of New Zealand by granting legal personhood to the Moon.[37]

Rather than simply asking how we'll determine who owns which piece of property on Mars or the Moon, let us instead ask what we want our relationship to that property—and to each other—to be. Is space a treasure chest, ready to spill its riches to the first group who can gather the funding and initiative to show up and take them? Is it the "province of all mankind," as described by the Outer Space Treaty, to be shared by all humans in the spirit of exploration and cooperation?[38] Regardless of how we decide to use space today, some of our descendants may eventually call it home, passing little plots of extraterrestrial land from parent to child the way humans have on Earth for most of our history. One day there might even be a human who will be able to point to a spot on the map of outer space and call it their birthplace. The legal bonds to space property that we're beginning to invent now will eventually be joined by emotional ties. An insignificant patch of rock on the Moon or Mars that's barely visible to us today might one day become the place where someone met their first love, or taught their firstborn how to walk, or even, eventually, where their ancestors are buried.

5

HOW WILL WE SHARE THE SPACE ENVIRONMENT?

The year is 1996 CE, and you are a coal miner in the Appalachian Mountains of West Virginia. Well, you used to be a coal miner, anyway. These days, the coal companies have started getting at the coal by exploding the tops off mountains, and there aren't as many regular mining jobs to go around. Coal mining saw your granddaddy through the Great Depression, but now it looks like you're out of the family business. Even though you're no longer working in the mine, it seems like coal dust is everywhere, ever since the blasting began on the nearby mountains: in the air, in the water, on your hungry children's faces. It's like all this money is raining from the sky, but to you it's as worthless as dirt.

The year is 2015 CE, and your small house in the Marshall Islands is flooding again. The latest king tide has brought a surge of dirty saltwater through the makeshift seawall you'd constructed around your property. Your government tells you that the encroaching seas are a result of the warming caused by the burning of oil and gas for fuel. Even now, your leaders are overseas, asking the rest of the world for their help in protecting your home, and promising that the Marshall Islands will reach zero emissions by 2050. But you know that your country's contributions to climate change are only a drop in the ocean. If the big polluters refuse to listen to your leaders' pleas for help, what will become of you and your family as your island is washed away by the tides?

The year is 2100 CE, and you find yourself with a free day in your space settlement, so you strap on your pressure suit and go for a hike. You head for your favorite spot: the top of a crater ridge, overlooking a fantastic view. But when you get up there, you see a sign posted: Development Coming Soon. A manufacturing waste disposal company has bought the rights to dump toxic waste here. Soon they'll be tearing up your beautiful view, and you won't be able to come anywhere near it without risking radiation exposure. You had hoped to take your children up here someday, but now they will never know it. Is there anything you can do? Or will you just have to find another favorite hiking spot? What happens when more and more of the nearby surface gets taken over by waste disposal, mining, and new habitats?

_____

MOTIVATION

NEW WORLDS TO EXPLOIT AND DESTROY

One of the greatest appeals of space travel for private spaceflight companies is the opportunity to access, extract, and sell new resources. Space entrepreneurs anticipate vast fortunes to be made from asteroid mining, space solar power, or microgravity manufacturing, once there's enough infrastructure in place to make these industries profitable. But despite the infinite size of outer space itself, the resources within our reach are not limitless, and the unsustainable use of those resources could cause long-lasting or even permanent damage to the space environment, ruining it for future generations. This is not a new scenario for humans: On Earth, we've eaten entire species into extinction, turned forests into deserts, and filled the oceans with our plastic waste. Environmental science has helped us understand, in horrifying detail, exactly what kind of damage we are continuing to inflict on our planet today, but as a civilization, we have not yet solved the problem of how to balance our other interests and desires with the need to protect the environment from ourselves.

Even if space resources *were* infinite, unequal access to those resources will only exacerbate the inequalities in human society. Again, this is a lesson we've already learned on Earth, where the environmental damage wreaked by anthropogenic climate change is, even now, disproportionately harming the poorest members of society, who have contributed the least to carbon emissions.[1] Will an increased use of space actually improve humanity's future, or will it enrich a small portion of our society while the rest are left behind on a deteriorating Earth?

We may not have solved our environmental problems here on Earth yet, but we are at least finally coming to the realization that the unchecked pursuit of resources can lead to irreversible environmental damage and increasing inequality. Now is our chance, as we prepare to live and work in the new environment of space, to figure out how to apply these hard-won lessons of sustainability and environmental justice beyond our planet. How can we ensure that the new resources we access in space will benefit all humans, including our neighbors on Earth and our descendants in the future? What do we owe each other and the generations of humans to come?

LESSONS FROM THE OIL INDUSTRY ON EARTH

PETROLEUM: A BLESSING AND A CURSE

Gaining access to a valuable natural resource, either by discovering a new source or by developing a new technology that depends on the resource, can bring both positive and negative effects to a community. There is perhaps no better example of this duality on Earth than fossil fuels, especially petroleum. Crude oil, natural gas, and coal are vital sources of energy across the globe, powering manufacturing, agriculture, transportation, and other vital industries. But these fuels take millions of years to form, and we're pulling them out of the ground much, much faster than they can be replaced. Our limited supply of fossil fuels is also highly localized; to make money from oil or gas extraction, a corporation needs both extraction technology and ownership or use rights for a deposit. This tends to concentrate the opportunity for wealth and increase tensions over land use rights.

The use of fossil fuels for energy has made modern civilization possible, raised the quality of life for billions of people, and made a much smaller subset of people very, very rich. But there are darker sides to the industry. Economists have noted that the presence of extremely valuable mineral resources, including fossil fuels, can counterintuitively slow economic growth and worsen the political outlook of a country, compared with nations without these resources. The existence of petroleum reserves in a country, for example, is correlated with strengthened authoritarianism, heightened corruption, and a larger number of violent internal conflicts.[2] Often called the "resource curse," the exact causes of this phenomenon can be difficult to disentangle from other factors affecting a country's economy and government. But the hard work of identifying how countries can escape the resource curse through better management of their mineral reserves could help us avoid falling into the same trap in the future.

Mineral resources in space will share many of the same features as petroleum reserves on Earth. For example, political scientist Michael L. Ross argues that four distinctive qualities of the petroleum industry enhance the resource curse on Earth: the incredibly large scale of oil revenues, the funding of a country's wealth from oil assets rather than taxes, the secrecy surrounding oil transactions, and the instability of the oil market.[3] Some of these qualities will be difficult to avoid in space. Ross notes that oil price instability, for example, is partly caused by the large upfront costs of petroleum extraction, preventing oil producers from reacting quickly to changes in demand. The upfront costs of asteroid mining will be even higher, potentially leading to similarly slow reactions in the future asteroid resource market. But by identifying these potential causes of the resource curse in the petroleum industry, we can adapt our plans for the space mining industry accordingly. For example, private companies and governments could work together now to develop techniques for regulating the future space mineral market, to try to avoid the volatility that the oil market experiences today.

Over the past century, we've also come to realize that the extraction, transportation, and use of fossil fuels can cause enormous damage to our environment. But the distribution of risk and harm is not equitable. The people who contribute the least to fossil fuel usage are often also the most vulnerable to the harmful effects of climate change. The US and China are the largest producers of carbon emissions, for example, but it is smaller, poorer countries like the island nations of Tuvalu, the Maldives, and the Marshall Islands whose territories are literally disappearing beneath the rising sea levels caused by those emissions. Our future descendants will also have to live with the consequences of our obsession with fossil fuels; not only will there be fewer accessible oil reserves left for future generations to use but the damage that we've caused to the Earth's environment in the form of oil spills, deforestation, and climate change will outlive us, leaving our descendants with all of the disadvantages of the oil industry and none of the benefits.

Resource extraction in space will likely not pose the same kinds of risk to the Earth's environment, but there will be humans living and working in space, sharing an environment with the machinery and byproducts of the space mining industry. Will we repeat this same pattern in space, where the poorest and least powerful residents of a space settlement will have to endure the harmful effects of a deteriorating environment while the mining industry leaders enjoy cleaner, safer habitats elsewhere in space or even back on Earth? Will future generations of humans in space struggle to live in a scarred, toxic landscape after years of unregulated mining, manufacturing, and waste disposal?

ORBITAL DEBRIS AND OVERCROWDING

POLLUTING OUR OWN BACKYARD

Humans have already begun to extend our troubled relationship with our environment into space. One of the clearest examples can be found close to home, in Earth's orbit. Our planet is surrounded by artificial satellites relaying communications, taking scientific observations, conducting aerial espionage, and helping commuters navigate around traffic jams. Satellites provide a startling array of services to our tech-hungry population, but they also represent a decades-long failure to plan for the future.

According to the European Space Agency (ESA), about 9,600 satellites have been launched into orbit around the Earth since the launch of the first artificial satellite in 1957. As of February 2020, about 5,500 of those satellites are still in orbit, but only around 2,300 are still functioning.[4] The rest have simply been left in orbit, either because it was cheaper than actively deorbiting them at the end of their mission lifetime (which would require launching them with extra fuel), or because they stopped responding due to equipment failures. Eventually, atmospheric drag may cause some of these defunct objects to fall to Earth, but in the meantime, they continue to orbit out of our control.

Intact satellites, operational or otherwise, are not the only artificial objects in Earth's orbit. Our skies are also filled with debris: discarded rocket boosters, leaked coolant, flecks of paint, equipment dropped by spacewalking astronauts, and fragments from accidental collisions or explosions. The ESA estimates that there are 34,000 pieces of debris larger than 10 centimeters in size, 900,000 pieces between 1 and 10 centimeters, and 128 million between 1 millimeter and 1 centimeter.[5] These tiny bits of junk are whizzing around the Earth at orbital speeds of up to 15 kilometers per second, ten times the speed of a bullet.[6] Impacts between debris and operational satellites or spacecraft can result in anything from a chipped window to a catastrophic collision. But given the prohibitive expense of cleaning up this debris, the problem has continued to worsen over the decades as we keep launching technology into orbit.

Article VII of the Outer Space Treaty states that the launching state is liable for damage caused by their spacecraft.[7] For example, if a rocket launched by Australia fails and crashes into a house in California, the Australian government is liable for the damage. Similarly, if Australia's rocket collides with one of the US's satellites in orbit, Australia is liable for the damage to the American satellite. But this legal liability doesn't solve the overall problem of orbital debris. For one thing, it can be very difficult to tell who a colliding object belongs to, especially if it has already been smashed to pieces by a previous collision. Secondly, liability requires the launching state to pay for damage to other spacecraft, but this payment does not change the fact that a new debris field was created by the collision. While financial liability can be a useful mechanism for motivating extra efforts to prevent accidents, such as encouraging oil companies to improve their shipping methods, it does not cure the environmental damage once an accident has occurred. Liability doesn't magically erase oil spills on Earth, and it won't clean up the debris in Earth orbit either.

Astrophysicist Donald Kessler famously warned in 1978 that the orbital debris problem could intensify out of our control in the event of a collisional cascade, where the fragments from a single spacecraft collision would cause a second generation of collisions, whose fragments cause further collisions, eventually producing fragments faster than atmospheric drag can remove them. In this scenario, even if no more satellites were launched, Earth orbit would be so full of fragments that it would become unusable.[8] This catastrophic result of an exponentially growing cascade of collisions, known as Kessler syndrome, was popularized more recently by the 2013 sci-fi movie *Gravity*, in which the planned destruction of a Russian satellite sets off a chain reaction of collisions that destroys several occupied spacecraft and countless communication satellites. In the decades since it was proposed, astrophysicists who study space debris, including Kessler himself, have debated the statistical likelihood of a Kessler cascade and the speed at which it would occur. This is complicated by the technological challenge of tracking smaller pieces of debris, suggesting that we might not recognize the cascade in its slower, early stages.

"The thing that is inarguable is that there is a finite carrying capacity to orbital space," aerospace engineer Moriba Jah said. "And that that capacity can be exceeded."[9] Jah is an associate professor of aerospace engineering and engineering mechanics at the University of Texas at Austin and has been working to better understand the orbital debris problem in order to find solutions. Jah likes to use the model of a "carrying capacity" because of its clear parallel to the concept of a carrying capacity in a terrestrial environmental context. The carrying capacity of a region on Earth represents the maximum population of a given species that could be supported by that region's food supply and other resources. Estimates of the Earth's carrying capacity for humans, for example, play a key role in predictions for overpopulation, and climate change is likely decreasing this carrying capacity.

By connecting orbital overcrowding to terrestrial environmental science via the idea of carrying capacity, Jah and other space scientists can harness people's established understanding of resource overuse in a context they may not have considered before. This is especially useful when discussing an environment like Earth's orbit, where there is no wildlife being threatened, no humans except the inhabitants of the International Space Station, and not even a landscape to damage. It is literally empty space. In fact, the only threat that humans pose to the environment of Earth's orbit is the threat to our own activities. On the other hand, the orbital space around our planet stands between us and the rest of the solar system. If the accumulation of debris makes this space unusable, we will be shutting off our only route to a multi-planet civilization. Addressing the threat that we pose to this intangible environment is vital for safeguarding our hopes for a future in space.

How close are we to exceeding the carrying capacity of Earth's orbit? This is a very difficult question to answer because the space debris problem is characterized by uncertainty and a significant lack of data: We don't know how many pieces of debris are in orbit. We don't even know the exact orbital positions of many of the objects we are tracking, which makes it difficult to predict where they'll be at a given time in the future. Besides addressing this fundamental shortage of data, we also need to be able to quantify the carrying capacity, so we can measure how rapidly we are approaching it. For example, the carrying capacity could be estimated by considering the collision probability, the potential severity of fragmentation, and the lifetime of every object in Earth orbit.[10] Jah has his own simple, functional definition of the orbital carrying capacity: "We know that it's exceeded when our decisions and our actions can no longer prevent loss, destruction, or degradation of space services and activities. If the things we do can't prevent these things from happening, then by all intents and purposes, that carrying capacity has been exceeded."[11]

Quantifying the orbital carrying capacity in order to figure out how close we are to exceeding it is a crucial first step in addressing the orbital debris problem, but Jah explained that there are additional benefits to establishing a definition for the carrying capacity. "Right now, it's very clear that there's the equivalent of a 'land grab' in space," he said. "Countries have the right to just launch however many things that they want. There's no global governance on that sort of thing." But if there were an internationally agreed-upon definition of our orbital carrying capacity, Jah predicted, "Then I think we can have an international discussion to say, 'How is humanity going to manage this common capacity?' Because the capacity does not belong to any given state; it belongs to humanity. So how do we, as humanity, manage this capacity?" This represents a shift in our perspective of these orbits: from a pile of valuable resources up for grabs for whomever can get there first, to a communal resource to be allocated in a way that maximizes the benefit for humanity. Instead of allowing nations or corporations to permanently occupy certain orbits—or at least for the decades that it would take their satellites to fall out of orbit—Jah suggested a model where "when we license somebody to operate in space, we're actually leasing some of that capacity to Acme, Inc., with the intent that Acme, Inc. will have to give that capacity back at some point," by intentionally deorbiting their satellite, for example.

Jah's team is also working on another metric for orbital sustainability that he calls the "space traffic footprint," "which would be like a carbon footprint analogue that could be loosely understood as the burden that any given object poses on the safety and sustainability of anything else," he explained. "So, assign it to objects dead and alive. That would create a marketplace, in many senses, because then you could put a bounty on objects based on the footprint it has." Jah suggested that space insurance companies could also charge premiums based on the predicted footprint that a satellite is expected to take up. These kinds of financial incentives could be very helpful for companies that are already working on technologies for cleaning up space debris, like a "chaser" spacecraft that would grab a single large piece of debris and then deorbit with it, or sticky nets to sweep up clouds of smaller fragments.[12]

It's vital that we figure out a good system of space debris removal and mitigation soon, and not just because the accumulation of debris could leave us trapped on our own planet. We have also already begun reproducing this problem around other worlds, with scientific satellites orbiting the Moon and Mars. There are so few objects in these orbits today that they pose little risk of collision to each other, but that will likely change during human settlement, when the demand for communication, navigation, and reconnaissance satellites will greatly increase. Which makes now the perfect time to transfer the lessons we've learned on Earth into space, instead of simply repeating the mistakes of the last several decades of orbital pollution. The costly cure we face for overcrowding in Earth orbit could provide the motivation we need to implement an ounce of prevention around other worlds.

The first step is to take the problem seriously. Space lawyer Chris Newman noted that, as with the gradual, early effects of climate change, it's difficult to galvanize people to care about environmental destruction that they can't see in their day-to-day lives. "People can't look up and see the skies are clouded," he said. "People can't look up and see the dangers from hundreds of thousands of small fragments, so it doesn't exist. So, there's an advocacy job to be done, as well as a regulatory job."[13]

SPACE MINING AND RESOURCE OVERUSE

LEARNING TO SHARE WITH EACH OTHER AND WITH FUTURE GENERATIONS

Certain orbits around the Earth are more valuable than others, due to the advantageous positioning that they provide. For example, a satellite in geostationary orbit, a circular orbit roughly 35,000 kilometers over the Earth's equator, will maintain its position in the sky over a particular place on Earth's surface at all times, making it much easier to track the satellite from the ground. This is especially useful for weather, navigation, and communication satellites, but there is a limit to how many satellites can be packed into geostationary orbit before they begin to interfere with each other's radio frequencies. Today, the allocation of geostationary orbital slots is managed by the United Nations' International Telecommunication Union, but in 1976, eight countries along Earth's equator attempted to claim sovereignty of the slices of geostationary orbit over each country's territory.[14] This claim has been broadly rejected by the majority of other nations; the main point of contention involves whether geostationary orbit should be considered part of space (which would make it immune to claims of sovereignty under Article II of the Outer Space Treaty) or an extension of the Earth.

However, the attempt to claim sovereignty over geostationary orbit reflects a larger worry on the part of developing nations without space programs (including those eight equatorial countries): If the use of space resources, like geostationary orbit, is allocated on a first-come, first-served basis, as is the usual model for orbital slots and radio frequencies, then by the time these nations develop their own spacefaring capabilities, won't other countries have already used up or even damaged the most valuable parts of space?[15] How can we reconcile this likely scenario with the Outer Space Treaty's requirement that activities in space "shall be carried out for the benefit and in the interests of all countries, irrespective of their degree of economic or scientific development," and that space "shall be free for exploration and use by all States without discrimination of any kind, on a basis of equality"?[16]

This is a problem not just for orbital real estate but also for more tangible space resources, like the water and precious metals that space mining companies hope to extract from asteroids, the Moon, and Mars. "We're talking about exhaustible resources, as well," noted space ethicist James Schwartz, a philosopher at Wichita State University. "Not just limited resources but exhaustible ones . . . These are the kind of things that are going to get used up pretty quickly. And once they're gone, it's not as though somebody can just hop in at the next slot and do everything they want to do."[17]

This is not the narrative presented by space mining companies as they attempt to convince investors and the public of the potential profit to be made by their industry. Planetary Resources founder Peter Diamandis has promised that the solar system holds "literally near-infinite quantities" of "the things we fight wars over," including metals, minerals, fuel, and energy sources.[18] But astrophysicist Martin Elvis estimates that only about ten near-Earth objects are rich enough in rare-Earth metals to be worth the cost of mining, and only about eighteen have enough water.[19] Objects in the main asteroid belt will be more plentiful, but even more expensive to reach. Elvis and his colleagues have also argued that while valuable resources like water, iron, uranium, and precious metals do exist on the Moon, they are heavily concentrated in certain areas, increasing the potential for conflict among different groups hoping to extract those resources for profit.[20] The Moon's surface also includes small areas whose special features, like near-constant sunlight, make them likely sites for future disputes over territory or use rights.[21]

"We'll have piracy and rustling and claim-jumping and espionage, all going on [in space]," Elvis predicted when asked what kind of conflicts over space resources he anticipates. "And some of it is illegal. Piracy is illegal, because you're coming alongside somebody else's spacecraft and taking the gold from it, or the platinum, or whatever."[22] The biggest challenges in fighting piracy will be prevention and enforcement. On the other hand, Elvis pointed out, "Espionage [in space] is actually encouraged. The Outer Space Treaty says you must be open to inspection, with reasonable notice, and so on." Article XII of the Outer Space Treaty requires "All stations, installations, equipment and space vehicles on the moon and other celestial bodies" to be open to representatives of other treaty signatories, with "reasonable advanced notice."[23] So, if a mining company strikes a huge vein of iron on the Moon, for example, their competitor could easily uncover this stroke of luck during a routine and perfectly legal "inspection" of the mining facility.

Let's take a step back from the potential for conflicts between individual mining groups and consider the use of space resources by humanity as a species. Just as a slot in Earth's orbit could be used for commercial telecommunications, scientific weather observations, military surveillance, GPS navigation, or even human habitation (but not all of these things at once!), a single patch of land on Mars or the Moon could be useful for extracting water, building habitats, collecting solar power, searching for signs of life, or building telescopes. But these activities cannot all be simultaneously performed in the same place, so how should we prioritize them when allocating resources as a species?

"I'm not as concerned about issues with fair access as I am concerned with issues of fair use, here," philosopher James Schwartz said, "Because when you have extremely limited resources, now it becomes even more of an ethical decision about, 'What are these resources going to be used for?' Are they just going to satisfy some really rich person's desire to have a hotel on the surface of the Moon? Or is it going to do something like contribute to valuable human scientific research?"[24] This echoes Moriba Jah's concept of orbital carrying capacity as a resource that belongs to humanity as a whole and which should therefore be managed by humanity as a whole.

Allowing space resources to be extracted on a purely first-come, first-served basis won't give us the opportunity to make these decisions, yet this is exactly how current space law allocates use rights. Article IX of the Outer Space Treaty states that if the activities of one state "would cause potentially harmful interference with activities of other States," the first state "shall undertake appropriate international consultation before proceeding with any such activity or experiment."[25] Martin Elvis explained that this article would help protect fragile instrumentation from what he calls the "very dirty business" of mining a planet or moon with little to no atmosphere: "You'll have tons of dust and stuff going not into the air, but into the vacuum. And it'll just come down on its little parabolic arcs. [It] could be a long way away; [it] depends on the dust size, and how energetic the mining machinery is. So you could well coat an operating telescope with dust, and that would not be good for it."[26] But under the Outer Space Treaty, if the telescope were built first, the prospective miners would have to engage in "appropriate international consultation" before building their dust-spewing mine.

On the other hand, Elvis noted that this same law could be used to intentionally prevent states from accessing valuable resources. "You could deliberately put a telescope in the center of a resource-rich region," he explained. Once it's built, "you open it up and say, 'Oh, is this a special place? I had no idea.' And then nobody can land within several kilometers, because they will destroy the value of your investment." More specifically, the telescope operators would be able to "request consultation" from anyone planning to land or mine nearby, although it's unclear yet what the legal consequences of such consultation would be.

Allocation of space resources will not just affect the humans squabbling over them at the time. Many of these resources are non-renewable, like oil on Earth. If we mine all the iron ore out of the Moon's surface, we are effectively taking it out of the hands of future generations of humans living on the Moon. On the other hand, if we transform that iron into structures like habitats, farms, manufacturing plants, or even schools and libraries, we will be improving the lives of our descendants. How should we decide what portion of nonrenewable space resources we can ethically consume in our lifetimes, and what portion should be preserved for later use?

INTERGENERATIONAL JUSTICE IN SPACE

OUR OBLIGATIONS TO THE FUTURE

The vast majority of people who will be affected by our decisions in space have not yet been born. This is true on Earth, as well, and yet we find it difficult to keep these unseen, unborn people in mind as we go about our lives. This is understandable, considering the urgency and tangibility of the problems facing our society today, compared with the vague and hypothetical struggles that future generations will face. As former US vice president and environmental activist Al Gore has noted, "The future whispers while the present shouts."[27] But most ethical systems hold that we have a responsibility toward these future generations. For example, there will be more of them than there are of us (assuming that the extinction of the human species is not in our near future), so a utilitarian approach would note that improving the lives of our uncountable descendants is worth some amount of work and sacrifice by the present generation. As philosopher Roman Krznaric writes, choosing to disregard the welfare of future generations "would be to display an unashamedly colonial attitude to the future, treating it as a distant land empty of inhabitants that we are free to plunder with impunity."[28]

Most human cultures believe they have a duty to fulfill their intergenerational obligations. The people of the Iroquois Confederacy of North America, for example, consider the effects their decisions will have on their descendants up to the seventh generation. According to Oren Lyons, faithkeeper of the Onondaga Nation of the Iroquois Confederacy, "We are looking ahead, as is one of the first mandates given us as chiefs, to make sure every decision that we make relates to the welfare and well-being of the seventh generation to come, and that is the basis by which we make decisions in council."[29] How can we protect the rights and quality of life of not only the first generation of human settlers in space but all the generations that will come afterward?

The pursuit of intergenerational justice is hindered by our inability to ask future generations what they want from us. One approach to overcoming this obstacle is to use philosopher John Rawls's "veil of ignorance": imagine that we have been tasked with determining the allocation and preservation of a nonrenewable space resource, like water on Mars, but that once we've made our decisions, we will be dropped into a random position in time and space and will have to live with the consequences.[30] For example, we might be miners living on Mars two hundred years from now, or we might be living in urban poverty on Earth today. We might be the first generation of landowners on the Moon, or the children of third-generation asteroid miners. How would this lack of knowledge about our destiny affect the choices we make about space settlement? This thought experiment forces us to consider intergenerational equity in our decisions, and to balance our resource use between the present and the future.

Environmental lawyer Edith Brown Weiss argues that one principle of sustainability and intergenerational equity should be "the conservation of options," or "conserving the diversity of the natural and cultural resource base, so that each generation does not unduly restrict the options available to future generations in solving their problems and satisfying their own values."[31] Brown Weiss also suggests that decision-making groups like governments should include a representative of future generations, a sort of "guardian *ad litum*" to argue for our unborn descendants' rights and needs in their absence.[32] In 2007, for example, Hungary appointed a Parliamentary Commissioner for Future Generations (later merged into the broader Commissioner for Fundamental Rights in 2012), and similar proposals have been made for the appointment of a United Nations High Commissioner for Future Generations at the international level.[33] In the early days of space settlement, when there will be constant conflict between short-term profits and the long-term sustainability of a settlement, appointing a representative of future settlement residents could help remind the settlement's decision makers that the resources of space must be shared among multiple generations of humanity.

ENVIRONMENTAL JUSTICE AND SUSTAINABILITY

LIVING WITHIN THE DOUGHNUT

Our obligations to future generations suggest that a key goal of a society's policies and attitudes should be environmental sustainability, an approach in which the current generation tries to live in such a way that their descendants will be able to enjoy the same quality of life. This perspective differs from the growth mindset fueling much of today's space settlement enthusiasm, which seeks to always increase the size of a society over time, in terms of population, territory, economic strength, scientific knowledge, technological development, and so on. If space is infinite, after all, and its resources limitless, why should we constrain ourselves to a certain level of resource consumption?

The first, obvious answer is that space resources are not infinite. More specifically, we can only use the resources that we can reach, and for now our technological abilities are limited to our solar system. Beyond this quibble, building a sustainable society requires more than just maintaining a balance between the rates of resource usage and extraction. After all, a dead civilization is technically sustainable, in that it does not consume more resources than it produces. But we have higher standards for human civilization, both on Earth and in space. We want our societies to flourish and our children to lead happy, fulfilling lives, but we also want to avoid destroying the environment in the process.

Economist Kate Raworth brings together these two requirements for a healthy society in her "doughnut economics" model. In this framework, the goal for a society is to maintain its position within a doughnut-shaped region bounded on the outer edge by an environmental ceiling, representing the limits of the physical environment, and on the inner edge by the social foundation, representing the obligation to provide its population with basic human needs. To thrive, a society must keep its members from plunging through the floor into poverty and deprivation while simultaneously avoiding breaking through the ceiling with environmental damage and resource overuse.[34]

Raworth's work encourages policymakers and economists to move away from the pursuit of economic growth for its own sake and instead to work toward bringing society into the "safe and just space" within the doughnut. The obsessive focus on economic growth that Raworth discourages is characteristic of Western, capitalist societies, but there are other worldviews that seek to balance ecological and social health that predate the doughnut economics model.

For example, Kyle Whyte, a professor at the University of Michigan, studies Indigenous perspectives on environmental and climate justice. "Indigenous traditions focus a lot on the idea that actually, what matters the most for sustainability is whether the relationships within a society are ones that are characterized by particular types of values and particular types of moral bonds," he explained. "So, there's really no such thing as a society that has a zero-carbon footprint and is deeply patriarchal or deeply colonial, because even if it's true that its carbon footprint is zero, then there's likely people that had to be displaced, like Native people, to do so, or people that are living in horrible household situations with dominering male figures, or it's a society that's deeply ableist, for example."[35]

If our goal is to build better societies in space, we'll need to do more than simply ensure that we can build habitats with a steady supply of breathable air and drinkable water. To be truly sustainable, space settlements will need to address both the physical limitations of their harsh but fragile environments and the ethical responsibilities to the health and well-being of their citizens. As meteorologist Eric Holthaus writes regarding humanity's current struggles with climate change, "In this moment of transformational change, we need to start by asking foundational questions, like, What is a good life, and how can it be possible for everyone?"[36] These same questions will serve us well as we design new communities in space.

HOW CAN WE PROTECT OUR DESCENDANTS FROM OURSELVES?

REINING IN OUR HUNGER FOR RESOURCES

Various legal structures have been proposed for off-Earth environmental conservation. For example, Article 7 of the Moon Treaty requires states to report "areas of the moon having special scientific interest in order that, without prejudice to the rights of other States Parties, consideration may be given to the designation of such areas as international scientific preserves for which special protective arrangements are to be agreed upon in consultation with the competent bodies of the United Nations."[37] While the Moon Treaty ultimately failed to gain widespread acceptance, others have also suggested designating scientific, environmental, and historical preserves in space. Charles Cockell and Gerda Horneck propose designating "planetary parks" on Mars, the Moon, and other planetary bodies.[38] These parks would function like national parks on Earth, protecting unique or significant parts of the landscape from development.[39] Similarly, Daniel Capper recommends designating separate types of environmental protection areas on Mars: parks for "human recreation," reserves for "sustainable industry and science," and an "Ecosphere Reserve" that "will allow only transient human presences to undertake no-footprint science, and otherwise here Mars will be left to Mars."[40]

Even in the absence of an international space environmentalism treaty, individual nations could regulate the activities of their private spaceflight companies. Space ethicist James Schwartz would like to see scientists be more involved in the regulation of private spaceflight and space mining companies, to help determine whether certain regions of another world should be preserved for scientific study. "You know, when I've said something like this in front of [private space] industry people, they get shocked, like, 'How dare you?!'" they said in an interview. "But if you think about it, some oversight would be helpful to industry."[41]

Schwartz explained that a scientist combing through the data about a potential mining site to determine whether it needs to be preserved will be generating valuable information for mining purposes. "If you've got scientists going through some material, they're actually going to give you a catalogue of what's there, ultimately. And so I think a very good partnership with the space industry on this score would be important, not only for increasing the amount of good science that's done, but for safeguarding some of it."

Space debris expert Moriba Jah suggested that if we want to prevent repeating our history of environmental destruction on new planets, we should learn from Indigenous societies who have managed their environments sustainably on Earth. Most importantly, he says, we need to acknowledge that different groups will seek to use space in different ways, but that we can find common ground by focusing on which values we want to carry with us into space. "Have a platform of ethics and values that you can honor and respect," Jah said. "Different people might have some differing values, so come up with some norms of behavior that you can agree upon in a shared resource."[42]

Martin Elvis, along with space ethicist Tony Milligan, has proposed a framework for protecting future generations from overshooting their resource usage in space due to unconstrained exponential growth. Elvis and Milligan recommend that as a rule, human development should be limited to one-eighth of the solar system's resources, and that this limit should be agreed upon long before it is reached, so that humanity has time to develop sustainable activities in space.[43] However, as Elvis admitted, it's unclear how to implement and enforce such a rule when the one-eighth threshold will probably not be reached for several centuries.

"What institutions last three hundred years?" Elvis said. "Not many. What laws last three hundred years? Well, the U.S. Constitution might make it! Who knows?" He crossed his fingers hopefully. "There are some universities that last that long. There are various religions that last that long. But it's very hard to create a new religion, [or] the equivalent of a religion, that would last that length of time. So I don't know. I think it's a big challenge."[44]

Humans—indeed, every living species—alter their environments to suit their needs. We will undoubtably alter the space environment even more than we already have as we take up residence beyond our planet. But that doesn't mean we can't make a deliberate effort to develop a sustainable relationship with our new environment, and to pass on to our descendants not only a stockpile of space resources for their own use but also a culture of sustainability and environmental justice. At the end of a long day of mining, farming, exploration, child-rearing, habitat-building, or whatever else they'll find to do, space settlers will hopefully be able to take a moment to find a quiet place to sit, away from the noise of imported human civilization, and just appreciate the view.

6

HOW CAN WE PROTECT THE SPACE ENVIRONMENT?

The year is 1639 CE, and you are a French trader from the colony of Quebec. You're glad to be on the road, enjoying the late spring air. You'd been ill back home, where the pox has been troubling your neighbors, but you're starting to get your strength back. In the distance, you see the smoke from a Huron village. Good people, the Huron, and excellent trading partners. Now that you think of it, you've never seen smallpox scars marring the faces of the Huron people. You worry briefly about bringing the pestilence to their village. But you've got goods to trade, and several more stops before your journey home. And surely if they haven't suffered the pox by now, God must have some reason for protecting them, right?

The year is 1859 CE, and you are a wealthy English settler in Australia. You could do with a spot of rabbit hunting, but although settlers have been bringing rabbits to the continent since the First Fleet, they're still hard to come by in the wild. So, you send for one or two dozen rabbits from England, which you release on your large property. As long as you give them a bit of time to breed first, you should be able to maintain a nice stock of game for the occasional hunting party. You find yourself looking forward to the first hunt. After all, it's just a few rabbits . . . what harm could they do?

The year is 2100 CE, and your space settlement wants to expand into a new region of the surface. But the surveyors have just reported back, and there's a problem. They've discovered evidence of microbial life in an underground pool near the proposed construction site. The biologists are already arguing with each other about whether this is truly extraterrestrial life, or just an invasive bacterial species brought here from Earth by the settlers. Meanwhile, the engineers admit that the power plant that was supposed to be built in the area will almost certainly be harmful to this tiny ecosystem. Should you give the scientists time to study or perhaps transplant the newly discovered life before continuing construction? Should you avoid building anything that might damage the extraterrestrial microbes? How much should the settlement limit its own activities to protect this indigenous, but microscopic, life?

––––––––––

MOTIVATION

BROADENING OUR ETHICAL PERSPECTIVE

The previous two chapters have focused on humans and our relationship with each other: How will we share the space environment, and the resources that we find there? What are our ethical responsibilities to each other? This is an anthropocentric view of space, focused on humans as both the decision makers and the potential victims of those decisions. But what about our ethical responsibilities to the space environment itself?

Approaches to environmental ethics, both on Earth and in space, lie along a spectrum: on one end, anthropocentrism prioritizes the needs of humans and assigns value to an environment based only on what it can provide for humanity. For example, in an anthropocentric perspective, species are worth protecting from extinction if they provide food or other useful raw materials, or if they play an important role in an ecosystem that provides such resources. Otherwise, they have no value in an anthropocentric system of ethics. A lifeless planet like Mars could be useful as a pristine laboratory for geological or astrobiological study, a cornucopia of new resources, or as the home of a future branch of human civilization, but it has no value beyond these kinds of human-serving purposes. Conflicts over the use of space can be resolved under anthropocentrism by determining which actions would best serve humankind.

However, many environmental ethicists and activists expand their definition of "entities who have rights" to include non-human animals in addition to humans, often weighted by some measure of each species' intelligence or capacity for self-awareness. Such an expanded view, for example, would argue against the harvesting of intelligent species like dolphins for food. This can be extrapolated even further to a biocentric perspective, which considers all life to have intrinsic value, independent of its utility for humans.

A move toward biocentrism complicates the ethical questions related to space settlement. For example, if we operate under the agreement that all life has intrinsic value, does that mean that every individual being—every human, dog, tree, worm, and microbe, both terrestrial and otherwise—have the *same* value? While forms of this argument have been put forth by environmental ethicists like Albert Schweitzer or religions like Jainism, living within such an ethical system makes it very difficult to, say, disinfect a bathroom, let alone decide how to use the space environment. More commonly, ethical systems tend to prioritize the rights of various species based on their capacity for reason, or, alternatively, their ability to experience pain.

No individual life form can survive for long on its own, because we all participate in larger, multi-species ecosystems. Ecocentric perspectives of environmental ethics find value not only in individual living beings but in systems of life. For example, an ecocentric perspective on space settlement might consider the value of expanding Earth's ecosphere into space, rather than the expansion of human civilization alone. While ecocentric ethical systems are common in Indigenous societies, in Western philosophy they emerged later, perhaps most famously in Aldo Leopold's "land ethic," which states, "A thing is right when it tends to preserve the integrity, stability, and beauty of the biotic community. It is wrong when it tends otherwise."[1]

If we shift to a wider, less anthropomorphic view of space settlement, we add a new dimension to our consideration of space environmental ethics, and are left with new questions: How do we weigh the rights and values of terrestrial and extraterrestrial environments with our human plans for space? What responsibilities do we have to the life that might exist on the worlds that we hope to visit, explore, inhabit, and mine?

OUR HISTORY OF ENVIRONMENTAL DESTRUCTION

HUMANS: THE GREATEST THREAT TO LIFE ON EARTH

"A bad smell of extinction follows *homo sapiens* around the world," writes author Ronald Wright.[2] Indeed, humanity's expansion across the surface of our planet has been a story of nonstop ecosystem destruction and species extinction. As our population has grown, so has our hunger not only for food but also for the fur, feathers, fiber, and other resources that plant and animal species can provide. Overharvesting has significantly contributed to several famous historical extinctions, including the North American passenger pigeon, which was hunted in huge numbers for meat and sport by European settlers over the course of the nineteenth century, and the great auk of the North Atlantic, prized for its meat, eggs, and down feathers until its extinction in the mid-nineteenth century.[3] Today, overharvesting threatens a third of the world's fisheries, as well as individual species like the Chinese pangolin, which is threatened primarily by illegal poaching for its scales.[4]

Humans have even engaged in the occasional intentional effort to eradicate various species. Most famously, the smallpox and rinderpest viruses were eradicated in humans and cattle, respectively, after decades-long global vaccination efforts. The elimination of deadly disease has not been the sole motivation for historical attempts to eradicate individual species, however. The American bison was brought to near extinction by overhunting in the nineteenth century, motivated partly by the US Army's attempts to deprive Indigenous people of their primary food supply and drive them off their land and onto reservations.[5]

The vast majority of human-caused extinctions, however, have been triggered indirectly, by the destruction of habitats.[6] Modern civilization depends on the extraction of resources from land that previously harbored thriving ecosystems. Some of this land is used for housing, replacing the habitats of countless other species with our own, but most is used not for occupation but rather to supply our other needs: mining, logging, waste disposal, agriculture, livestock grazing, and so on. We divert rivers to power our dams and blast apart mountains to access coal seams. Anthropogenic climate change even causes harm to ecosystems that until now were relatively untouched by human use.

Humans have also indirectly triggered species extinctions and ecosystem damage via the introduction of invasive species. Intercontinental species mixing reached a peak during the era of European colonization, as colonizers carried crops, livestock, and unnoticed stowaways like rats and earthworms across the world.[7] One of the clearest examples is the introduction of European rabbits to Australia. While the First Fleet of settlers brought captive rabbits with them as a food supply, it wasn't until the mid-nineteenth century that breeding populations appeared in the wild, after landowning settlers began releasing imported rabbits on their properties in order to build up a stock of game for hunting. With no natural predators (until foxes and feral cats were also introduced), the rabbits multiplied, their population soaring into the millions as they spread across southern Australia. The rabbits' grazing habits devastated both native and sown pastures, leaving sparse, weed-infested fields in their wake, and their tendency to feed on seedlings inhibited the growth of native trees and shrubs. Several small mammal species that depend on this native vegetation for food have disappeared since the rabbit population began to spread.[8] Australians have attempted to control the rabbit problem via hunting, warren destruction, extensive fencing, and even the intentional introduction of viruses deadly to rabbits, but the infestation remains.

If there is life in space, human settlement could threaten its existence through our actions there, just as we threaten the diverse communities of life on Earth. It's unlikely that we'll overharvest extraterrestrial life (unless it turns out to be edible, which seems unlikely), but we will be introducing invasive species into the local ecosystem, intentionally or otherwise, as we grow crops, dispose of human waste, and build self-sustaining communities. As we spread our civilization into space, we'll also be altering the space environment to suit our needs, which could very easily pose the threat of habitat destruction to any indigenous life. Can we learn from our destructive past and avoid causing further environmental damage and extinctions in space?

PLANETARY PROTECTION

KEEPING THE COSMOS CLEAN

The mere presence of terrestrial technology has the potential to cause harm to any life that exists in a space environment, even before any humans set foot there. Article IX of the Outer Space Treaty requires that as we explore the Moon and other celestial bodies, we should "avoid their harmful contamination and also adverse changes in the environment of the Earth resulting from the introduction of extraterrestrial matter."[9] Space agencies have developed various "planetary protection" policies and techniques to prevent possible contamination of the Earth by materials from space, or of the space environment by terrestrial organisms. These two types of interplanetary contamination are known as "back" and "forward" contamination, respectively.

Back contamination represents possible "adverse changes" to the Earth's environment from "extraterrestrial matter." This would include infection by an alien microbe, the star of many science fiction horror stories. Even outside of fiction, there are plenty of cautionary tales on Earth regarding invasive species or communicable diseases crossing borders, causing harm by their mere presence in their new environment. Most notably, the accidental introduction of new diseases like influenza and smallpox by European explorers in North and South America killed up to 90 percent of the Indigenous population, wiping out entire societies and changing the course of history for the hemisphere.[10] The potential for back contamination from space was accounted for during the Apollo missions, in which human astronauts as well as lunar samples returned to Earth from the Moon. Astronauts from the Apollo 11, 12, and 14 missions were quarantined for three weeks after their return, along with the samples they had collected, despite the confidence of scientists at the time that the threat of a lunar plague was extremely unlikely.[11]

Article IX of the Outer Space Treaty also covers forward contamination, the potential "harmful contamination" of the space environment due to the presence of terrestrial matter. The parts of the solar system that we have explored so far are extremely unwelcoming to life, but we already know of terrestrial organisms, often called "extremophiles," that thrive in harsh environments like highly acidic volcanic springs, the frozen ice of

Antarctica, and the deepest parts of the ocean floor. If some of these extremophiles hitched a ride to another planet on a lander or an astronaut's space suit, they might find a way to survive and even evolve to thrive in their new home before we knew what was happening.

Margaret Race, a biologist who works with NASA on problems of planetary protection, stated that a major concern since the early days of space exploration has been the threat that forward contamination could pose to the integrity of astrobiological research. "Surprisingly, at the time of Sputnik, back in the late 1950s, a group of biologists were already saying: If we can send spacecraft out there to the Moon and beyond to look for life, we don't want to spend these millions and millions of dollars (now billions) and get up there and discover life but find that it was launched with our spacecraft from Earth."[12]

Even within the scientific community, conflicts might soon arise between the opportunity to improve our astrobiological observations in space and the possibility of contaminating the very environment we're trying to study. Space ethicist Kelly Smith of Clemson University described such a scenario: "Suppose that there's a warm, salty, ocean underneath the ice cap on Europa, and we have reason to suspect that there's life down there, like we've sampled some of the steam coming off the vents, and there are cells." This could suggest the presence of a rich, complex underwater ecosystem beneath the ice cap. How could we study this ecosystem? "There's absolutely no way you can put a probe down into the Europan ocean without contaminating it. It's just not possible," Smith said. "So, if the point of preserving life is to preserve the science, but you can't do the science at all without taking some risk, now you've got a real quandary."[13]

Since 1959, the United Nations' Committee on Space Research (COSPAR) has provided recommendations for preventing forward and back contamination during space missions. COSPAR categorizes destinations in space based on their vulnerability for contamination and suggests different procedures for each category. For example, a mission to Mercury requires no planetary protection efforts, due to Mercury's obvious past and present uninhabitability, while a trip to Venus or the Moon has only a "remote chance" of contaminating future scientific studies there, and merely requires some pre- and post-mission documentation. However, a mission to a planet or moon vulnerable to "significant contamination" of its habitable regions must include extra precautions: the spacecraft should be constructed in a cleanroom, for example, and if the craft will actually touch down on the surface of another world, it may need to be sterilized prior to launch.[14]

Of course, the planetary protection problem gets a lot more complicated when astronauts are included on a mission. Humans can't be sterilized; not only would most techniques used for sterilizing spacecraft kill a human astronaut but we now know that a healthy human body depends on a complex internal microbiome of bacteria and other organisms. As Race acknowledged, "If humans go [to Mars], we know that there's no way we can put them in totally enclosed systems. So what do we do about the contamination? Do you essentially autoclave everything that goes out [of the habitat]? Is that even possible? How do you clean and maintain spacesuits? How do you keep the lab on the surface separate from the habitation area where the astronauts might sleep and work?"[15] This problem will worsen as humans move toward permanent habitation in space. How can we grow crops, process waste, and raise children without exposing the surrounding environment to our microbiomes?

Today's planetary protection policies and regulations are fairly anthropocentric, because they're primarily concerned with the scientific utility of extraterrestrial life rather than its intrinsic value. What about the potential harm that we might inflict on these hypothetical space microbes? Invasive species from Earth could do more than simply confuse future astrobiologists. They might outcompete native organisms, breeding like Australia's rabbits and causing our first interplanetary extinctions. How should we address this possibility in our plans for human space settlement?

Smith argued that there's a difference, ethically, between killing a few microbes and wiping out an entire species, especially a unique or unusual species. "Maybe you shouldn't locate your human settlement in a particularly delicate, rich ecosystem because you're going to be destroying microbes that will never be seen again," he suggested. On the other hand, "if there's a particular species of microbe which is ubiquitous on Mars, it's everywhere, then there's less of a concern, in the same way that there's less of a concern on Earth if building a settlement chases off several field mice. We don't really spend a whole lot of time worried about that, whereas [displacing] snail darters or some exotic species of zebra or something like that, that's much more concerning."[16]

Planetary protection rules will continue to regulate space activities in these early days when there are still so many unknowns about the potential for life in space, but if those questions are not answered quickly enough by astrobiologists, other space industries may grow impatient. As Mark Lupisella and John Logsdon write, "It isn't clear that scientific value will be enough to warrant the kind of conservative approach that may be needed to ensure the preservation of possible indigenous extraterrestrial life, thereby realizing that scientific value . . . Looking further ahead, we might also wish to consider how we will guide our actions when the scientific novelty wears off."[17]

Planetary protection regulations are under increasing threat as more and more profit-motivated (rather than science-motivated) organizations set their eyes on space. In 2019, the Israeli space company SpaceIL attempted to land the first privately funded spacecraft on the Moon. The lander, called Beresheet, was carrying a digital library created by the Arch Mission Foundation, a US nonprofit hoping to create an off-world backup of human civilization. But Beresheet crashed into the Moon, scattering its payload—including the Arch Mission Foundation's library—across the landing site. Only after the crash did the Arch Mission Foundation reveal that they had added several thousand tardigrades to the library.[18] Also known as water bears, tardigrades are microorganisms that can survive exposure to the vacuum, cold, and radiation of space.

Even if the tardigrades survived the crash, they likely remained in the dormant state they use to survive extreme environments, and there is nowhere on the Moon where they could rehydrate and begin to reproduce. This is exactly why spacecraft headed for the Moon are not required to be sterilized by COSPAR's planetary protection policy. On the other hand, COSPAR does recommend that mission planners document an inventory of all organic materials present on a lander bound for the Moon. The Arch Mission Foundation flouted this policy, according to cofounder Nova Spivack, by smuggling dormant tardigrades into the library aboard the Beresheet without informing SpaceIL. "Space agencies don't like last-minute changes," Spivack said in a later interview. "So we just decided to take the risk."[19]

Asking forgiveness rather than permission will be a tempting strategy for future private space companies hoping to avoid the financial cost of following planetary protection rules. They will find allies in space settlement advocates like Robert Zubrin, who bristles at the very existence of planetary protection regulations as an overreach by scientists: "It's not just a matter of who gave the Moon to astrobiologists, but also of who gave the universe to professional scientists. Humans do not exist to serve scientific research. Scientific research exists to serve humanity," Zubrin writes.[20] As space continues to fill with people who prioritize profit or colonization over scientific exploration, our window to detect potential life in the solar system without terrestrial contamination is likely closing.

TERRAFORMING

MAKING SPACE HABITABLE

Exploration, habitation, and resource extraction all carry a risk of inflicting environmental damage in space, just as they do here on Earth. But some futurists and space settlement enthusiasts have proposed an even more drastic alteration of the space environment: the transformation of the surface of a planet or moon into a more Earth-like environment via a process known as terraforming.

The atmospheric chemistry, pressure, and temperature inside an artificial space habitat is, by design, Earth-like enough to be habitable by humans, but it requires enclosure by pressurized walls and constant maintenance. Terraforming would affect the entire surface of a planet, rather than just a smaller "indoor" region, and by planetary scientist Christopher McKay's definition, the environment of a terraformed planet "must be stable over long time scales and must require no, or a minimum of, continued technological intervention."[21] After an initial input of energy and effort, a terraformed environment would behave like Earth's natural environment and essentially maintain itself.

For example, in 1961, Carl Sagan speculated on the possibility of the "microbiological re-engineering" of Venus by introducing blue-green algae into its atmosphere. The algae would use photosynthesis to convert the planet's abundant carbon dioxide into oxygen, which would also reduce the greenhouse effect and lower Venus's surface temperature.[22] Sagan later turned his attention to the potential for "re-engineering" Mars, a planet now considered to be one of our best candidates for successful terraformation. Mars has the opposite problem as Venus: instead of harboring a thick, toxic atmosphere with a runaway greenhouse effect maintaining deathly high temperatures and pressures at the surface, Mars lost nearly its entire original atmosphere to solar wind, leaving surface pressures so low that liquid water cannot exist. To terraform Mars, planetary engineers would need to increase its surface temperature and atmospheric pressure while protecting the atmosphere from solar wind. Sagan suggested spreading a dark material, or even growing dark-colored plants, on Mars's polar ice caps, allowing them to absorb more of the Sun's heat, increasing the surface temperature while releasing water vapor and carbon dioxide into the atmosphere.[23] Other researchers have explored the feasibility of importing greenhouse gases or building giant orbital mirrors to increase Mars's surface temperature, constructing a magnetic shield to protect Mars's atmosphere, and releasing genetically engineered microbes onto the planet's surface to alter the atmospheric and surface chemistry.[24]

Terraformation is the ultimate example of long-term planning, as even optimistic estimates predict that it would take centuries of effort and patience before a human could walk unprotected on the surface of Mars. Advocates of terraforming Mars or other space environments see it as a crucial step toward creating a truly multi-planet civilization. Zubrin even claims that the successful terraforming of Mars would demonstrate humanity's superiority over the physical world: "The first astronauts to reach Mars will prove that the worlds of the heavens are *accessible* to human life. But if we can terraform Mars, it will show that the worlds of the heavens themselves are *subject* to the human intelligent will."[25]

Whenever someone waxes poetic about humankind bending the universe to our will, it's worth taking a moment to consider the ethical implications of the proposal. One major consideration about terraforming is that the process could damage or even wipe out any existing life on the planet being terraformed. If an alien microbe evolved on Mars, it probably would not survive in a more Earth-like environment, so by transforming Mars's surface into Earth's, we might exterminate species or entire ecosystems without even detecting their existence. The changes we would make to a cold, dry, relatively airless world like Mars would also introduce physical processes—such as wind, flowing water, and new chemical reactions—that could easily erase or contaminate any evidence that extraterrestrial life ever existed on the surface. If we allow planetary engineering to race ahead of astrobiological research, we could miss our opportunity to make what would be the most important scientific discovery in human history: the discovery of life that evolved beyond our planet. We also risk exterminating the very lifeforms we dream of discovering.

The ethical dilemma of terraforming far exceeds planetary protection concerns about forward contamination by a lander or even a human settlement. The goal of terraforming is to intentionally create an entire ecosystem on a global scale, which would more than likely destroy any existing ecosystem. Terraforming technology might even become feasible before we definitively determine whether extraterrestrial life exists on the planet or moon that we hope to transform. But suppose we do discover evidence of existing microbial life on a planet like Mars. Should this disqualify Mars as a target for terraforming? Should we avoid settling on Mars at all?

Carl Sagan famously argued for exactly this stance: "If there is life on Mars, I believe we should do nothing with Mars. Mars then belongs to the Martians, even if the Martians are only microbes. The existence of an independent biology on a nearby planet is a treasure beyond assessing, and the preservation of that life must, I think, supersede any other possible use of Mars."[26] Christopher McKay even argues that if microbial life is discovered on Mars, humans should not simply leave Mars to the microbes, we should "undertake the technological activity that will enhance the survival of any indigenous Martian biota and promote global changes on Mars that will allow for maximizing the richness and diversity of these Martian life forms."[27] In other words, we should engineer the surface of Mars not to improve its habitability for terrestrial life, but for Martian life!

Space ethicist Kelly Smith finds these types of arguments, that humans should avoid worlds where microbial life might already exist, difficult to defend. "You have to first grant that microbes, as a class of organisms, are somehow on the same level with human beings," he said. "I'm not saying you can't make an argument to that effect, but it really stretches credulity. It's an uphill battle."[28] After all, humans have already demonstrated that we are willing to intentionally eradicate disease-causing viruses like smallpox to prevent human death and suffering. Admittedly, viruses are not unequivocally considered to be "alive," and there were some ethical concerns during the development of the vaccine that smallpox eradication represented a "new form of genocide."[29] But given the opportunity, humans would likely jump at the chance to exterminate deadly microbial species like the bacterium that causes cholera or the parasite that causes malaria. Unlike these terrestrial microbes, however, hypothetical Martian microbes currently pose no danger to humanity, or even to individual humans. They may merely someday stand in the way of our off-Earth expansion. Space settlement advocates argue that such an expansion is vital for humanity's long-term survival, but does this potential for indirect harm justify their extinction?

It may seem premature to debate the ethics of using a technology that does not yet exist to indirectly destroy an ecosystem that may not exist at all. But our potential for inadvertently exterminating a unique species or ecosystem in space might arise long before we develop the technology to terraform entire planets. By the time we come to an agreement about the ethics of terraforming and planetary protection, it might be too late.

COSMOCENTRISM

THE VALUE OF A LIFELESS WILDERNESS

Suppose we could determine absolutely that no life existed in a certain space environment, like the surface of Mars, that we wanted to terraform. Even in this scenario, the process of terraforming would still permanently change the environment of the planet. Freshly melted water would flow through the landscape, beginning the long process of erosion and covering low-lying rock features. Plants and microbes imported from Earth would spring up, covering Mars's characteristic red surface with Sagan's proposed blue-green algae. Even the color of the sky would change as the atmosphere thickened and its chemical composition slowly evolved. By the time the project was complete, the formerly stark and lifeless view would be replaced with a cacophony of plant life and human technology. We would, in effect, be destroying the original environment of the world we choose to terraform. This won't be the kind of destruction that the environment can ever recover from, as a forest on Earth might eventually recover from overlogging or a natural disaster. Long after humans have gone extinct, the scars we leave on other worlds will remain. Is this ethical?

Our helpful extrapolation of terrestrial environmental ethics into space falters here, given that this field of ethics has until recently focused exclusively on an environment where life is abundant. Dilemmas regarding the protection of any part of the Earth's natural environment are deeply entwined with the protection of the local or global ecosystem. The value of a physical environment on Earth is almost always viewed through its relationship to life, or at least the potential for life.

"If you look at most views of environmental ethics," Schwartz said, "the thought that there's some ecosystem there that ought to be preserved is a real guiding thought. Maybe it's not an ecosystem, maybe the focus is on the species level, or the individual organism level, but the presence of life has been very important in most views."[30] Does this mean that a truly lifeless environment has no intrinsic value, beyond the potential value for humans who want to terraform it? Or, having grown up in an environment teeming with life, have we simply not yet properly explored the potential value of an abiotic wilderness?

Even the word "wilderness" tends to call to mind an untamed ecosystem, like a forested mountain range, a murky jungle, or a lion-filled savanna. One of the major pieces of environmental legislation in the US, the Wilderness Act of 1964, defines wilderness as "an area where the earth and its community of life are untrammeled by man, where man himself is a visitor who does not remain."[31] Space beyond low Earth orbit is certainly an area "where man himself is a visitor who does not remain," at least for now. But if there are no communities of life in space for us to trammel, can it still be considered a wilderness, deserving of our protection?

Wilderness clearly has value to humans, even beyond the physical resources or scientific knowledge it can offer us. For example, wilderness can benefit individuals: American environmental writers and activists like Henry David Thoreau, Aldo Leopold, and John Muir observed that spending time in the wilderness can be a transformative experience for humans, a way to broaden our perspective by reconnecting us to the natural world.

This argument was used to advocate for wilderness protection legislation like the Wilderness Act of 1964, in parallel with more pragmatic, conservationist concerns about managing the nation's natural resources sustainably. The wilderness of space has an equal capacity for producing transformative experiences for humans traveling beyond the Earth. Holmes Rolston III writes that as we continue expanding our use of space environments, we should "preserve those places that radically transform perspective . . . For intellectual and moral growth one wants alien places that utterly renegotiate everything in native ranges."[32] Astronauts have already described experiencing similar shifts in their perspective while in space, often while looking back at the Earth.

These are still anthropocentric arguments, however. The broader question is whether a lifeless environment has any intrinsic value beyond what it can provide humanity or other forms of life. Physicist Martyn Fogg notes that beyond anthropocentrism and even biocentrism, there is a wider approach to environmental ethics that he calls "cosmic preservationism."[33] In this perspective, also known as cosmocentrism, even the lifeless components of nature have intrinsic value; the very existence and uniqueness of the rocky landscape of Mars or the Moon makes it worthy of protection. Philosopher Robert Sparrow, for example, claims that attempting to terraform another world would be an act of "arrogant vandalism" that would demonstrate "the sin of hubris."[34] Certainly, terraforming would forever alter the natural landscape of a space environment, removing its original beauty from existence. Sparrow argues that this decision reveals "an ethically significant aesthetic insensitivity."

Cosmocentrism is not a widely accepted view in space environmental ethics. "It becomes challenging when you've got an environment that you take to be completely lifeless," Schwartz said. "You have to sort of get a little creative in terms of the ethical argumentation that you provide. Maybe the claim is that something of great beauty is there and would be lost. Maybe the claim is something about how, even if there's no life there, it would still demonstrate this great hubris on our part to completely remake a world. Maybe we can describe planetary surfaces as just valuable in themselves. And I don't think any of these are incoherent claims, but I think they end up being fairly difficult claims to provide good justifications for."[35]

Even within a cosmocentric ethic, in which a lifeless landscape is considered to have some intrinsic value, it does not necessarily have the *same* value as an ecosystem. Proponents of terraforming usually take the biocentric position that, all things being equal, "life" is better than "no life," and that bringing a terrestrial ecosystem to a lifeless world would not only expand and protect Earth's ecosphere from existential threats but would also improve the formerly uninhabited planet or moon. Is this hubris? Perhaps. But vandalism seems like an extreme accusation, given that the goal of terraforming is not destruction for its own sake, but rather the replacement of the original landscape with another: the cultivation of a garden in a desert.

WHAT IS OUR RESPONSIBILITY TO THE SPACE ENVIRONMENT?

A GARDEN BEYOND THE EARTH

The question of whether to terraform another planet is far in our future, but planetary protection concerns are already relevant, and will become even more so in the near future as human activity in space increases. We have some policy and regulation in place to help protect the hypothetical extraterrestrial life we will encounter, but will those rules and guidelines survive the coming age of private resource extraction efforts in space? And what if we do detect life, or indications of life, elsewhere in the solar system? Would such a discovery halt human activity in that region, the way the discovery of a population of an endangered species will occasionally halt construction projects here on Earth? Or will stakeholders argue that the benefits of space activities for the human race outweighs the harm done to the local space ecosystem?

According to Kelly Smith, it's crucial that we work out a system of principles and values now, before specific questions arise in space. "Now that does get immediately really complicated," he admitted, "because the problem that you encounter in ethics is [that] unlike in science, consensus is rare." Ethicists don't even agree on the relative value of humans, intelligent animals, microbes, and a lifeless environment; what hope is there for the larger population of scientists, space entrepreneurs, government regulators, and the public? But Smith argued that the conversation itself is vital to prepare ourselves for a larger future in space, because it will at least allow us to establish our principles and maybe even find common ground. "But you won't really know until you get into the messy debate and you really try to hash all this out. So that's what I'm arguing for: Let's spend the next twenty years yelling at each other and trying to figure some of this stuff out. And then when we're actually in a situation to make a decision, maybe it'll be a bit clearer."[36]

In these ongoing conversations about space bioethics and planetary protection, perhaps we should even consider appointing advocates for the hypothetical extraterrestrial life that we hope to protect. Such advocates could speak for the Martian microbes during the planning processes for exploration, settlement, or mining missions to Mars, the way the appointed guardians of the Whanganui River in New Zealand defend the rights of the river in legal cases.

Eventually, in the more distant future, terraforming may become feasible enough that serious plans are proposed for converting Mars or another world into an environment more habitable for humans. If by then we have confirmed that no life exists on the world to be terraformed, we will still face lingering ethical questions about the intrinsic value of the lifeless landscape, not to mention the environmental justice concerns discussed in the previous chapter. How can we ensure that terraformation is conducted ethically and justly, for the environment we hope to transform as well as for humans both in space and on Earth?

In his master's thesis, Robert French proposed an ethical framework in which terraforming could be conducted in a sustainable and environmentally just way. French's model views the human species as a gardener whose responsibility is to guard and tend to a garden plot, the planet Earth. As humanity initiates plans to terraform a new world, the gardener "desires to expand his or her garden and after prepping a nearby plot begins to transplant seedlings from the first plot to the new one. The gardener's first concern is the health of the entire gardening endeavor, and only a foolish gardener would work the new plot at the expense of the old." French's ethical framework would call on humanity to continue to care for the Earth as enthusiastically as we would be creating a new ecosystem on the terraformed world, countering the "disposable planet mentality" that worries some opponents of space settlement. The concept of humans as a gardener also guides us to respect the intrinsic value of the environment we are shaping, in addition to the useful purpose it serves: "The gardener intimately cares whether or not the garden succeeds, both for the gardener's own sake, because of all the time and effort that the gardener has put into the garden, and for the plants themselves, because the gardener has nurtured these lives, coaxed them into being and supported their struggles to survive and flourish."[37]

French's model would likely not satisfy those who believe we have no right to alter an extraterrestrial landscape at all. But it does represent a way to counter a strictly anthropocentric and colonialist narrative in which humanity has the right to extract whatever resources we desire from space with no concern for the harm we may be doing to each other, the environment, or non-human life. Ecosystems are complex and messy, environmental ethics perhaps even more so. We will always be acting without complete information about the environments we'll encounter in space, and we may never reach consensus about the ethics of our actions. But just as astrobiologists are working to provide as much data as possible about the potential for life beyond Earth, we should continue to work toward a system of environmental ethics in space to help us minimize the damage we will do there. If our descendants do one day grow gardens in space, let's ensure that they bring only nourishment, not harm, to all life that exists there.

HOW WILL WE LIVE WITH EACH OTHER?

7

WHERE'S MY MONEY?

The year is 1868 CE, and you have traveled far from your home in rural China to take a job as a railroad worker in a foreign country. You and your fellow laborers have spent the winter setting explosive charges and digging through solid rock with hand tools to cut a path through these mountains. The money you earn is a great help to your family back home, but you know that the white men working the line are being paid more than the Chinese men. The message has been passed up and down the line: alone, you are powerless, but if you all stop working, they will have to listen. This railroad cannot be built without you.

The year is 2017 CE, and you have been out to sea for months. At first, you were overjoyed to be offered this job on a fishing trawler. You had spent the last of your savings to pay for your illegal passage into Thailand to find work, and you'd finally found hope that you could earn some money to send back to your children in Myanmar. But now you are trapped on this boat, and the captain refuses to tell you when you'll return to shore. Even when you do make landfall, you won't be allowed to leave until you pay off your debt, the captain tells you. You hadn't agreed to take on any debt, but what can you do now? It's not like you can swim home from here.

The year is 2100 CE, and you've done it: you sold everything you owned and moved to space. You're broke, but you're happy. You've unpacked your suitcase, introduced yourself to your new neighbors, and settled in. It's time to get to work. After all, you'll need to buy groceries and pay next month's rent, and you have your eye on a new pair of space boots. So, who's hiring? How will you make money in space? What if your employer refuses to let you go home from the asteroid you're mining? How will you survive if you lose your job, but can't afford a ticket back to Earth?

_____

MOTIVATION

WHO IS THE ECONOMY FOR?

The biggest open question for private space companies right now is not the *technical* feasibility of space mining, settlements, or tourism; it's the *economic* feasibility of the industry. We know how to get people into space, keep them alive there, and bring them back to Earth safely (with some souvenirs). But these activities are enormously expensive. Even with recent advances in private launch vehicles, which cut the cost per kilogram of cargo by one-third compared to the Space Shuttle program, the price tag to move a kilogram of supplies to the International Space Station in 2017 was around $100,000 USD.[1] Both private corporations and government agencies have to be able to demonstrate that the outcome of their work in space is worth the investment, for either shareholders or taxpayers, but corporations are expected to produce those returns as profit, while governments have more freedom to pursue purely scientific or even nationalist goals in space.

Why do entrepreneurs even bother investing in the space industry, a risky and unproven market with a huge initial cost barrier? Why not direct their talents toward safer industries here on Earth? As with any untapped market, while the start-up costs and risks are large, the potential reward is huge. Entrepreneurs and investors are gambling that the space industry will strike metaphorical gold. Space entrepreneur Peter Diamandis has predicted that "Earth's first trillionaire will be made in space," while astrophysicist Neil deGrasse Tyson more specifically predicted that humanity's first trillionaire will be "the person who exploits the natural resources on asteroids."[2] Other space enthusiasts, motivated not by profit but by their dreams for a permanent and sustainable human presence in space, recognize that space settlement will require a stable extraterrestrial economy.

Much has been written on the economic feasibility of space, and the analyses and opinion pieces will only accelerate as companies continue to search for profit beyond low Earth orbit. Profiles of industry leaders like Musk and Bezos praise them as visionaries or decry their unrealistic ambitions. A smaller but still significant portion of the discussion focuses on the customers and clients of the growing space industry, like the millionaires who will likely fund the early space tourism industry, or the public–private partnership between NASA and SpaceX to return launch capabilities to the US after the end of the Shuttle program. But hardly any attention has been given to a third component of the future space economy: the workers.

Certainly, a lot of space work can be done by robots, but human workers offer a more flexible, adaptable, and potentially (once there's a self-generating supply of them in space) cheaper option. After Elon Musk gave interviews describing his plans for a luxury spaceliner to transport settlers and his hopes for "everything from iron refiners to the first pizza joint" on Mars, author Nicole Dieker pointed out that all of these dreams require a working class, one that Musk has yet to mention.[3] "Musk is asking us to imagine being refinery owners and restauranteurs," Dieker wrote. "He's not asking us to imagine being refinery workers or dishwashers—but those people will be on Mars too. They'll have to be."[4] Will these people benefit from the potential profits of the future space economy, or will they be exploited to power it?

Let's consider the space economy from the perspective of a regular worker living in space. What will they receive in exchange for their labor? How much control will they have over their workplace conditions? What if they run out of money or lose their jobs? Historically, on Earth, a person who is unable or unwilling to sell their labor to an employer can attempt to "live off the land," which could mean farming, fishing, hunting, or even scavenging in an urban setting. This is not a guarantee of survival, of course, and has led to countless deaths by starvation, exposure, or disease. But in space, where even breathable air must be generated and maintained inside pressurized habitats, striking out on one's own will lead to a much quicker death. This means that workers will be in a particularly vulnerable position, far from the incredibly habitable environment of Earth and dependent on others for transportation and protection from the dangers of space.

History has shown us that the combination of potential profit and a vulnerable workforce can easily lead to labor exploitation, especially in a remote environment that's difficult, if not impossible, to monitor for abuse. If we focus only on the stability and value of the space industry, we could end up with a thriving and profitable space economy built on the backs of exploited workers. As economist Matthew Weinzierl pointed out in 2018, "Even an established, efficient space marketplace offers no guarantee that the pursuit of private priorities in space will serve the public or respect the public's ethical judgments."[5]

ECONOMICS IN HISTORICAL COLONIES

RAW MATERIALS AND UNREGULATED LABOR

The recent growth of private spaceflight companies and their interest in mining and colonization has a direct parallel in the history of colonization on Earth. The European colonization of North America was not a purely governmental effort, but instead featured public–private partnerships, and was heavily motivated by the potential for profit.

"For-profit elements of colonization were baked in at the beginning," said Margaret Newell, colonial historian and professor of Early American History at the Ohio State University. "The colonization itself was conducted by private companies. The first corporations in modern history essentially developed as trading companies, and many of these trading companies were put in charge of colonization."[6]

These trading companies, and the European governments that partnered with them, were particularly interested in the natural resources available in the Americas. "When you look at the earliest descriptions of the New World, they read like commodity lists, in many ways," Newell said. "What people were seeing were things that were either already proven market values, tradeable commodities, or things that were scarce." This class of "scarce" commodities often included resources that Europeans had already depleted on their own continent, Newell explained. For example, "for people in England, who were stripping their own forests of wood to make charcoal for the industry there, or for fuel for their homes . . . they were obsessed with the forest [in North America]."

The similarities to today's burgeoning space industry are clear: just as European trading companies wanted to harvest raw materials that were abundant in North America but scarce in Europe, space mining companies hope to extract precious metals (rare on Earth by definition) and water (not rare on Earth, but rare and extremely useful in space) and make a profit selling them to Earth or to other groups in space. In fact, descriptions of space are also beginning to sound like commodity lists, as entrepreneurs begin appraising individual asteroids based on their predicted composition, size, and ease of access.[7]

But there was something else that attracted private European corporations to North America, Newell pointed out, something less tangible than lumber or fur: a lack of protection for labor rights. "There's a lot of new research that talks about how labor is one of these areas [that drew companies to the colonies], that this lack of constraint and custom in these places outside of Europe was attractive," Newell said. "These are global companies and global interests, and are consciously adapting new labor practices aimed at extracting more labor without payment . . . They try to apply those things at home, too, and can't quite do it to the same extent."[8]

These companies had trouble implementing their exploitative labor practices in Europe because oversight, labor protections, and public perception were too restrictive. "I think people saw these new places, the New World, and maybe space in the future, as places where all the rules were off," Newell said, "and they could engage in kinds of activities that wouldn't be limited by custom and practice and expectations at home. And that could lead to exciting experimentation, or it could lead to naked exploitation."

Similarly, many entrepreneurs are excited about the opportunities for experimentation in a less-regulated "Wild West" space environment. Obviously, no one involved in the space industry has publicly praised space for being an economic arena particularly conducive to labor exploitation. (Not as of this writing, anyway.) But the industry has advocated for limited regulation, pointing to the internet as an example where low regulation allowed for "permissionless innovation," a phrase coined by George Mason University policy analyst Adam Thierer.[9] Eli Dourado, a fellow technology policy researcher formerly also at GMU, drew this parallel between the internet and space during his testimony before the US Congress's House Committee on Science, Space, and Technology in 2017. "As with other domains," Dourado said, "this freedom to experiment [in space] will result in mistakes and failures. Yet over the long run, permissionless innovation will result in faster progress and more robust solutions to policy problems than a precautionary regulatory mentality."[10]

Space settlement advocate Robert Zubrin also argues that the scarcity of labor in space will help to stimulate innovation. "Just as the labor shortage prevalent in colonial and nineteenth-century America drove the creation of 'Yankee ingenuity,'" he writes, "so the conditions of extreme labor shortage combined with a technological culture will tend to drive Martian ingenuity to produce wave after wave of invention in energy production, automation and robotics, biotechnology, and other areas."[11]

But technology is only one of the solutions European colonists applied to their need for labor. Slavery, indentured servitude, and other forms of labor abuse also proved to be cost-effective remedies. As history shows us, an unregulated, remote environment like space can breed the "exciting experimentation . . . or naked exploitation" that Newell described.[12] Even if we manage to avoid repeating the mistakes of our past, there are undoubtedly plenty more forms of labor exploitation that we haven't even invented yet. Is that the kind of innovation we want?

PAST, PRESENT, AND FUTURE LABOR EXPLOITATION

BREAKING THE CYCLE

Space settlement advocates like Zubrin see a bright future for workers in space, as they assume that the high cost of transportation will make labor scarce, giving workers more power in negotiations. "Indeed, it can be safely said that no commodity on twenty-first-century Mars will be more precious and more highly valued than human labor time," Zubrin writes in *The Case for Mars*. "Workers on Mars will be paid more and treated better than their counterparts on Earth, and education will be driven to a higher standard than ever seen on the home planet."[13] But workers were scarce on the tobacco plantations of the early American southern colonies, too. The high cost of transporting enslaved workers across the Atlantic did nothing to protect those workers once they reached the United States.

Chattel slavery is the obvious example that springs to mind when speculating on labor exploitation practices that we risk reinventing in space, but humans have practiced a dazzling variety of subtler labor rights abuses. For example, "company towns," in which a corporation houses its employees in a remote area near their worksite, can easily become environments that exploit workers even when they're off the clock. Company towns often evolved from mining or logging camps, although one of the most notorious in the US was built by the Pullman Palace Cars Company to house its railroad car manufacturers. In Pullman, Illinois, and many other company towns, the residents paid rent to and bought food and amenities from the very company they worked for. In some company towns, workers were paid not in cash but in scrip, which could only be used at the company store. This monopoly allowed the store to drive up prices while simultaneously keeping the workers dependent on the company. In 1893, Pullman reduced wages in response to an economic depression but did not reduce rent to compensate, leading to a massive strike the following year. The isolated nature of space settlements could easily lead to a reinvention of these company towns.

In fact, many labor exploitation systems have been created in response to challenges that we will re-encounter in space: dangerous working conditions, in remote locations, with lots of money on the line. For example, both space historians and entrepreneurs have made comparisons between the growth of the private space industry and the construction of the nineteenth-century US transcontinental railroad.[14] The railroad, built by private companies with government grants, connected both coasts of the nation, allowing faster transportation of goods and passengers and stimulating the expansion of towns and industries in previously remote regions. It also made the railroad companies extraordinary amounts of money, so it's an appealing parallel for private space companies, as well as for space settlement enthusiasts advocating for extraterrestrial infrastructure. But digging deeper into the metaphor, we find a cautionary tale beneath the promising business plan. The vast majority of the construction work on the railroad was completed by workers from China, who performed backbreaking manual labor under incredibly harsh weather conditions for less pay than their white counterparts. In June 1867, Chinese railroad workers organized a strike to demand better pay. After breaking the strike by cutting off food supplies, the railroad company grudgingly raised wages, but the dangerous working conditions continued. Railroad workers were killed and maimed in explosions and rockslides, swept away by avalanches in the Sierra Nevada mountains in winter, and felled by the intense summer heat of the Colorado Desert. Despite their invaluable contributions to the project, written accounts of this labor force are inconsistent and scarce, but historians estimate that as many as 2,000 of the roughly 20,000 Chinese railroad workers were killed working on the line.[15] Isolated, underpaid, and unappreciated workers in a deadly environment built the infrastructure that connected the US coasts; will we repeat this exploitative system in space?

"I find the idea of space labor absolutely terrifying, and I haven't stopped thinking about it since I read your email," labor rights activist Sarah Newell (no relation to Margaret Newell) said while discussing parallels between labor rights issues on Earth and potential future issues in space. Newell worked for the International Labor Rights Forum and worried that today's corporation-led drive toward space settlement could put workers at higher risk than a purely national effort. "The biggest issue we're going to face is if it's going to be a corporate town, as opposed to a government settlement, it's going to be difficult to impose, really, any laws, the same way [that] it's very difficult to impose laws on corporations down on Earth, with all the freedoms and exceptions they're given. So, I think we need to be really concerned about what labor frameworks will be in place by the time corporations take people up there."[16]

Ensuring that labor protection laws are in place is only half the battle. Laws don't mean much if they're not enforced, or worse, if the lawbreaking is not even noticed by a distant authority on another planet. Newell observed that this is a crucial challenge for labor rights even here on Earth. "I think most people really believe in the terms of a contract. People believe, 'If I have a contract, or I have an agreement, I can always leave, and always change my mind. [But] when you have oppressed populations, that's not always the case. There's not a way to make them put you on a ship and take you home. Especially in the early days, at what point is a police force or security force or any sort of enforcers going to be installed, wherever in space you're settling?"

The challenge of enforcing laws in space extends far beyond labor protection and will be explored more broadly in the next chapter. But workers in space will be particularly vulnerable to abuse, given that they will likely rely on their employers for both transportation and life support in their workplaces. Newell described a clear parallel on Earth from her work with the ILRF: "The campaign we work on that I think is the closest is with migrant workers from Myanmar who work in the fishing industry in Thailand." Recruiters demand a fee and the workers' passports in exchange for transporting the workers to Thailand, Newell explained. The passportless workers are then informed that they owe the cost of their transport to the fishing company and must work off the debt. "And the real problem with the sea is that they can take these folks out on boats and keep them . . . and the boat simply does not come back to port, and there's no way to escape. Which makes me think of space: If you can't get a seat on the one shuttle or rocket that's going back, you're stuck with whatever your situation is."

Not only are these migrant workers in Thailand trapped on boats at sea with no way to return home, but they are consequently at the mercy of their employers—or rather, their captors. "There's a lot of rampant labor abuse on these boats," Newell said, "because if you have someone who you consider disposable, who you're not paying anyway, someone who doesn't have any ability to report you, what's your incentive not to work them as hard as you can to get the most money?"

GOING BROKE

HOW TO BE POOR IN SPACE

Now that we've explored employment in space, let's consider unemployment. What will be the fate of a space resident who does not work, either because they cannot or because they choose not to? The answer depends on the political and economic structure of the space civilization, but for now let's imagine a settlement with a purely market economy, in which there are no public goods or services, and every commodity must be purchased. Consider a worker from Earth who is offered a job off-planet, moves out to a space settlement, and then loses their job. If they have no other source of income, such as dividends from investments, or rental properties back on Earth, and cannot afford the cost of transportation back to Earth, the now-unemployed worker will be trapped in space with no funds. Even with a job, wages may be so low or prices so high that a worker could end up unable to afford even the basic necessities.

On Earth, this situation can easily lead to homelessness. In space, where humans will live in pressurized, climate-controlled, radiation-shielded habitats, living space will be much more limited, and therefore more expensive. What happens to a settlement resident who cannot pay for housing? Do they wander the halls of the habitat, setting up temporary shelters in unused corners? Perhaps homelessness will be criminalized, as it has been in many cities in the US, and homeless settlers will be arrested and punished by the settlement's criminal justice system.

Poverty on Earth also carries a high risk of malnutrition and starvation. We can ask the same questions about food in space: What happens to residents of a space settlement who cannot afford to buy food? Will we replicate in space the systems we maintain on Earth that allow people to starve to death while their richer neighbors continue to eat well?

Access to clean drinking water is another challenge faced by people living in poverty on Earth today. But while waterborne diseases are a much larger threat to this population than directly dying of thirst, in space, the lack of water itself (clean or otherwise) will be an even greater threat. There are no rivers in space, no oceans, no rain—nowhere to scavenge drinking water. Instead, water will have to be recycled in a closed system, as it is on the International Space Station today. Similarly, breathable air will also need to be recycled, scrubbed of carbon dioxide, and carefully monitored for contaminants. Not even the poorest residents of Earth have to worry about how to afford the air they need to breathe. The most urgent dangers for a poor resident in our imagined space settlement will not be starving to death or failing to find a place to sleep at night but losing access to the water dispensers or depleting their allotment of breathable air.

This hypothetical settlement, where poverty results in asphyxiation, represents one extreme on the economic spectrum. It seems more likely, if we extrapolate today's Western societies into the future, that space settlements will combine a free market with some kind of social safety net. The extent of this safety net will depend on the values and politics of the settlement's population: will health care be free or subsidized? What about housing, or transportation back to Earth? Practical concerns will likely also influence these decisions. For example, because water and air must be recycled in space, maintaining a stable closed system for each will be easier if there is a central authority controlling usage and distribution, rather than competing companies selling their services to individuals. For breathable air in particular, it would be extremely difficult to enforce a payment system: How would you shut down services to delinquent customers if they were walking around in the same habitats with residents who were up to date on their air bills?

As we work toward building a stable economy in space while ensuring protections for workers, we also need to consider the failure mode of the space economy: unemployment and, potentially, poverty. How will we protect all of the inhabitants of a space settlement, whether or not they are participating in the economy? Will members of the community have value beyond the labor that they can contribute?

WILL WE HAVE POVERTY IN SPACE?

BUILDING A SPACE ECONOMY FOR EVERYONE

The previous section imagined a purely capitalist space society, then considered a settlement with a mixed economy with a social safety net to address the consequences of poverty. A space settlement could, of course, establish a non-capitalist economic structure based on its ideology and priorities. Today's off-Earth residents—the rotating crew of the ISS—practice a system within the station that resembles communism more than capitalism. No money changes hands between crew members. There are personal belongings on board but no part of the station is privately owned by individual residents. Indeed, ISS crew members live a very Marxist lifestyle, contributing their labor "each according to his ability"—determined by their individual training and experience—and consuming food, water, and energy "each according to his needs"—based on their physical requirements.[17]

Of course, the ISS is not a self-sustaining civilization, and the community of astronauts and cosmonauts on board do not live in economic isolation from their (mostly capitalist) governments. Astronauts are paid for their work in space, and the supplies and materials sent to the ISS are paid for by taxes. But this framing suggests a question: As the first space settlers transition from their long journey to life in their new home, how will the circumstances of their voyage affect their economic choices? If the passengers and crew of the transport ship shared both resources and responsibilities equally among themselves, they may choose to simply continue this communal system in the settlement. Conversely, the new settlers might disembark from a luxury spaceliner with a built-in class divide between paying passengers and the stewards and cooks who pampered them during the journey.

Once again, the answers to these questions will be determined by the motivations of the settlers. Refugees fleeing a disaster on Earth will be more likely to emphasize a fair and equitable distribution of resources to maintain as high a survival rate as possible. Entrepreneurs and prospectors hoping to strike it rich in space will prefer a system that rewards risk-taking, even if it carries a high penalty for failure.

As the first wave of settlers considers their options and debates the merits of different economic systems, they will have to consider how their new economy will change and scale as their population increases over time. Will their chosen system tend toward stagnation or growth after a few generations? Will it be particularly vulnerable to instability or external disasters like widespread crop failures? Is it likely to lead to the concentration of wealth over time, creating stark inequalities between the living conditions of the rich and poor?

We cannot make these decisions for our descendants. But we can address inequality and labor exploitation in the space industry today. Private spaceflight companies have already faced accusations of labor rights violations. In 2017, SpaceX settled a lawsuit with employees claiming that the rocket company had not provided legally required meal and rest breaks or adequately compensated workers for overtime hours.[18] During the initial peak of the global coronavirus pandemic in 2020, employees at Blue Origin were "outraged" after they were pressured to travel from Washington state to rural Texas to conduct a test launch of a rocket designed for space tourism, potentially exposing them to the virus.[19]

Non-space companies like Tesla and Amazon—which are connected to SpaceX and Blue Origin through their CEOs, Elon Musk and Jeff Bezos, respectively—have experienced even larger labor controversies. In 2017, complaints surfaced from workers on the factory floor of Musk's electric car manufacturer, Tesla, alleging that dangerous and strenuous working conditions within the factory had resulted in over a hundred calls for emergency services since 2014.[20] A few years later, in 2019, Tesla was found guilty of violating the National Labor Relations Act for, among other things, Musk threatening his employees with the loss of stock options if they joined a union.[21] Jeff Bezos's online marketplace, Amazon, has also developed a reputation for "grueling and unsafe" working conditions, with injury rates at one Amazon warehouse in New York exceeding three times the industry average.[22]

Tesla and Amazon are not rocket companies, but they are headed by billionaires with publicly announced goals to settle space, and they may ultimately help fund this work. If we want to encourage our descendants to protect future space settlers from exploitation by their employers, we must set an example by ensuring that workers in today's space industry and related fields are safe, well-compensated, and empowered to negotiate.

Labor rights activist Sarah Newell agrees. "I think in the immediate term, the most useful thing we can do is fight for better, stronger standards, and better, stronger enforcement mechanisms on Earth," she said. "Because whatever systems companies adopt here, they're going to bring up there. And whatever solutions or international agreements we're able to form here, it makes it that much easier to say, 'Well, we've agreed to this standard for here. We should apply that in space.'" In particular, Newell argued, we should be "advocating for stronger standards here [on Earth], and supporting workers as they organize here, and supporting unions so that they can continue to organize. Because unions are going to be the protection in space."

8

WHAT IF SOMEONE STEALS MY STUFF?

———

The year is 2100 CE, and your new life in space is going great. You have a cozy subterranean home, an exciting and challenging job, and a great new pair of space boots that you're looking forward to breaking in during your next stroll on the surface. But today when you got home from work, you found that someone had broken in and stolen your new boots. You're pretty sure you know who did it—your neighbor has been talking about how great those boots are, and how much they wish they had a pair. Who do you report this to? Do you just want your stuff back, or do you want to see your neighbor punished, too? How can you keep this from happening again?

———

The year is 2100 CE, and your neighbor has just accused you of stealing their space boots. Unfortunately, your neighbor happens to be a prominent member of your space settlement, and you're just a recent immigrant from Earth who barely speaks the language. Things have moved quickly; your neighbor has no proof, but the rest of the settlement has been quick to believe their accusation anyway. Will anyone listen to your side of the story? How will you be punished if they decide you're guilty, if they haven't already?

———

The year is 2100 CE, and you're the leader of a space settlement that's just had its first reported crime. You were hoping to avoid having to fulfill your role as the settlement's judge; you've got enough on your plate already. The accuser is a popular member of the community, and public opinion seems to be on their side. But the accused has recently been recruited to the settlement because their expertise in hydroponic farming is desperately needed to address the latest crop failures. How will you determine what actually happened, and who's to blame? What's a fair sentence, and what's best for the safety of the settlement? How will you prevent similar crimes in the future?

MOTIVATION

BRINGING OUR SINISTER SIDE WITH US

In October 2018, an argument erupted between two men in the dining room of a Russian research station in Antarctica. Sergey Savitsky, an electrical engineer, and Oleg Beloguzov, a welder, had been working together for about six months, and their relationship was strained at best. Some reports later claimed that the argument was prompted by Beloguzov spoiling the endings of books that Savitsky had checked out of the station library, while others say that Beloguzov had been taunting Savitsky.[1] Regardless of the specific trigger, months of tension between the two finally boiled over, and Savitsky grabbed a knife and stabbed Beloguzov in the chest several times.

Beloguzov survived the attack, and Savitsky turned himself in to the manager of the research station and was detained in a church for eleven days until he was flown back to St. Petersburg, where he was charged with attempted murder, although the charges were later dropped with the support of a recovered Beloguzov. Some of the media reports of the incident blamed the psychological stress of life in Antarctica for Savitsky's violent outburst; others noted that the engineer had been drinking before the attack.

It's easy to imagine a similar scenario playing out in a space settlement. After months or years of living with the same set of people in a small, cramped settlement, surrounded by a cold, unforgiving environment, psychological stress and interpersonal conflict could escalate small disagreements into physical violence. But there is one key difference between the Antarctica stabbing and a hypothetical assault in a space settlement: in space, there is no flying back to St. Petersburg to stand trial. When violence or other crime occurs in space, the community will have to handle the consequences on its own.

A criminal act committed between individuals, the bread-and-butter of every police procedural and mystery novel, is likely the first scenario to come to mind when considering the concept of crime in space. It remains to be seen whether interpersonal violence is a permanent component of human nature, but no society on Earth has been able to rid themselves of it yet. Besides violence, there are other types of harm that individuals inflict on each other that communities tend to label as crimes. Theft, for example, violates the rules or customs of most societies, although the exact definition of "theft" varies, based on each society's conception of property rights.

An individual can also violate a society's rules by causing harm to the community as a whole. This type of offense, including negligence of duty, malingering, or outright sabotage, will pose a particularly serious threat to an early space settlement, where the labor of every individual will be crucial to the settlement's survival, and malicious damage to a fragile habitat or delicate equipment could wipe out the population. Beyond individual crimes, a community can also inflict harm on its members in a way that violates the community's values. Depending on its politics and laws, the society might not classify its own actions in this case as crimes, despite the damage they cause to portions of its population. Many of the grim futures hypothesized in this book fall into this category, which includes labor exploitation, institutional discrimination, and human rights violations.

Deviant, harmful behavior has plagued every human society on Earth, and each has developed its own system of justice in response, not to mention its own definition of justice itself. We will very likely carry this sinister side of ourselves with us into space. Which definition of justice will we bring along, and how will we choose to enforce it?

CRIME IN SPACE

CAN WE AVOID BRINGING IT WITH US?

Which actions are considered immoral differs between individuals, and which actions are illegal differs between governments. So, what kind of legal system will we use in space? There's no need to reinvent the wheel; we could just copy a terrestrial system into our space settlements. I proposed this scenario to space lawyer Chris Newman of the University of Sunderland. "Space tends to be a multinational enterprise," he countered. "So, the first question is, if we were [going to] do a cut-and-paste, whose laws would we cut and paste? Now, there's broad similarities . . . We don't allow murder, we don't allow stealing, we don't allow violation of the human body. . . . [There are] similarities. But also there's actually a substantial amount of differences, as well. So you'd first have to get agreement as to whose laws were going to be the operative ones."[2] We'll explore the challenge of developing systems of law and government beyond the Earth in the next chapter. For the purposes of this chapter, let's just assume that our hypothetical space settlement has a set of agreed-upon laws.

We can also think about crime beyond the context of a given legal system. Sociologists situate crime in the larger category of deviance, behavior outside the accepted norms of a society. When these norms are codified into law, the behavior becomes a crime, but control of deviant behavior is a problem faced by all societies, regardless of whether a formal legal system exists. Broadly, in this chapter, I will use "crime" to refer to an intentional act that violates the rules (formal or informal) of a community and inflicts harm on a member or members of that community. To minimize harm, space settlements will have to decide how they'll respond to such behavior.

The best method for minimizing the harm caused by illegal behavior is to prevent crime before it occurs. Prevention requires an understanding of the causes of crime, which has been an evolving field of research since the eighteenth century. Historical schools of criminology have argued, variously, that crime is a choice made by individuals pursuing their own self-interest, that a tendency toward crime is determined by an individual's biological or psychological makeup, or that structural inequalities like poverty create environments in which illegal behavior is considered necessary for survival. The benefit of identifying the causes of crime in a society—no easy task, given the complexity of the data—is that each theory suggests practical solutions for reducing crime. If crime is a rational choice, then crime rates can be decreased by increasing punishments, thus making illegal activities less appealing compared with the consequences of being caught. On the other hand, if criminal behavior has a biological or psychological component, such as mental illness or addiction, we may be able to identify and treat the conditions that predispose an individual to such behavior. And if crime is caused by social conditions, then addressing inequalities like poverty will have the added benefit of reducing crime rates.

RESPONSE AND JUDGMENT

LAW & ORDER: OUTER SPACE

Clearly, we have not eradicated poverty on Earth, nor have we solved the problem of crime prevention, so our terrestrial societies still require a way to respond to a crime after it has been reported. Organized law enforcement emerged with the first legal codes, and ancient empires experimented with police forces that performed duties such as apprehending thieves, maintaining public order, and guarding or recapturing the enslaved. The Roman Empire employed nightwatchmen who also acted as firefighters, a system that was later used in urban areas of medieval England and was then adopted by the American colonies. The world's first "modern" police force—professional, publicly funded, uniformed police who conduct preventative patrols—is usually identified as the Metropolitan Police Service of London, established in 1829.

Maintaining a dedicated police force in space will be extremely impractical in the early days, when population numbers (and thus crime) are low and labor demands are high. Perhaps initially the leader of the settlement will hold all police-related responsibilities, like the station managers in Antarctica. Later, a growing settlement may set up a volunteer watch, which could eventually be formalized into a professional force, paralleling the evolution of police organizations here on Earth. But this is not the only path available to us.

Many academics and activists have criticized the efficacy, safety, and fairness of our current law enforcement systems. In the US, the frequent killings of civilians by police officers, and police departments' disproportionate arrests of and violence against people of color (especially Black people), incited outrage building to widespread protests most recently in 2020. Common demands by protestors and organizers have included increased police accountability and oversight, an end to the militarization of police departments' weaponry and tactics, and a reduction of the scope of policing combined with diverting funding away from police and toward community support programs like public housing and addiction treatment centers.[3]

Other activists call for a more radical solution than reform or defunding: the complete abolition of police in our society. Alex Vitale points out in *The End of Policing* that many police departments have their origins in slave catching, strike busting, and protest suppression. "While the specific forms that policing takes have changed as the nature of inequality and the forms of resistance to it have shifted over time," Vitale writes, "the basic function of managing the poor, foreign, and nonwhite on behalf of a system of economic and political equality remains."[4] Vitale and other police abolitionists argue that not only do our current police organizations cause more harm than good but that this problem cannot be solved via reform, due to the very structure and ideology of the police as a group empowered to control behavior via violence.

Even for people who agree that the police should be abolished, it can be difficult to imagine how we would transition from our current justice system to a police-free society without further violence and instability. Space provides us with a unique opportunity to sidestep this problem by simply never developing police in the first place. After all, there are no cops in space today, and there will only be cops in space in the future if we choose to send them there. Instead, we could use the lessons learned from police reform and reduction on Earth to address crime and deviant behavior in space with other types of crisis response—like mental health intervention—and focus more effort on addressing the causes of crime.

Regardless of what kind of prevention programs or response teams we have, crime will likely still occur in space. The next step will be for the evidence to be presented to someone with the authority to decide whether the accused is responsible, and what should happen to them. Who will act as a judge in a space settlement?

In a tiny community like a transport ship or an early settlement, supporting inhabitants whose sole job is to provide legal judgment may be as untenable as supporting a police force, but on the other hand, it might otherwise be nearly impossible to find someone among the population who can be impartial. There simply won't be enough degrees of separation in the relationships between residents. As Newman pointed out, small, isolated communities like ships at sea or Antarctic research stations address this problem by giving the responsibility for judgment to people in positions of authority, who are presumably already practicing the emotional distance and impartiality required for effective leadership. But one scenario unique to space settlements may pose a particular challenge to efforts to pass unbiased judgment: what if the accused is the only person on the planet with certain expertise required for the settlement's survival?

It's not hard to imagine a scenario where the resident accused of a crime is also the only person who knows how to fix the water recyclers or is the community's leading expert in hydroponic agriculture. How would anyone whose life depends on these technologies be able to pass impartial judgment on this person? How can a judge fairly balance the accusations of a single victim against the safety of the entire settlement? The best solution to this problem is to ensure that it is avoided in the first place. Cross training and documentation should be a high priority for any space community, not just to protect against losing subject-matter experts to accidents or illness, but also to prevent those experts from holding the settlement hostage with the implicit (or even explicit) threat that they will withhold their desperately needed expertise.

CONSEQUENCES

THE PRACTICAL PROBLEMS WITH PRISONS

Now we come to perhaps the most difficult part of our journey through an off-Earth criminal justice system. Once the settlement has determined that one of its residents has broken the rules, what should they do with the rulebreaker? The standard response to criminal activity in much of the world, especially the US, is imprisonment. Proponents of the carceral system point to several useful functions of prisons: First, incarceration removes the perpetrator from society, at least temporarily, protecting the population from further criminal behavior. This loss of freedom is unpleasant, so the threat of prison can also deter potential offenders from committing crimes. Some proponents advocate for using prison as a rehabilitative environment, where inmates receive job training, mental health and addiction treatment, religious counseling, or other services meant to help them reintegrate into society after release and to prevent further criminal activity. Others support incarceration as a punitive practice, a way for society to enact a moderately humane form of revenge against people who have broken the social contract.

No prison system is guaranteed to serve any or all of these functions, of course. An incarcerated person entering prison for the first time may be able to learn a trade or receive counseling for anger management, but they may also be radicalized into white supremacy, taught advanced techniques for burglary, exposed to a new addiction, or connected to a network of organized crime. This does little to protect society or reduce recidivism. The evidence for the effectiveness of prison as a deterrent to crime at the population level is mixed, but at the individual level, the likelihood of reoffending is either unaffected or increased by the experience of prior incarceration.[5]

As transformative justice advocate and former New Mexico prisoner Justin Allen said, "Prison is violence. I don't care if you're in there for a day or ten years, you're going to experience violence or witness violence, in one shape, form, or fashion. It's a culture. That system is based on violence, and in order to survive in that environment, you have to assimilate to a violent culture." Allen found that the traumatic nature of prison made it difficult for him to break out of the cycle of substance abuse and reincarceration, despite the rehabilitation programs offered at the facilities where he was incarcerated. "I took every kind of class, everything I could while I was in there. And every time I got out, I kept repeating the same mistakes that put me there because some of the trauma that I endured while I was in there, I didn't deal with."[6]

Nonetheless, many futurists assume, if asked, that the carceral system will inevitably accompany us into space. Astrobiologist Charles Cockell has already suggested a design for a space prison that can operate on a planet's surface or in free space, called the Extraterrestrial Containment Facility, or Exoconfac.[7] Cockell describes various physical features of Exoconfac that would prevent escape and "harness the productive power of the inmates to the maximum extent possible under respectful and humane living conditions." But prison abolitionist and former prison librarian Courtney Montoya explained in an interview that even the architecture of terrestrial prisons helps to perpetuate trauma: "Even the way that the facilities are built: to keep people in extreme states of sensory deprivation, not having that human contact, seeing nothing but four walls in a very small space, not being able to be around nature, not being able to have enough access to sunlight, poorly made windows that aren't having sunlight come through . . . Those are all reproducing a state in peoples' psychological well-being and physical well-being that aren't treatment."[8] This already sounds like a description of many proposed space habitats for regular citizens, let alone carceral facilities. How much effort will habitat designers put into ensuring that space prisons are built with the psychological well-being of their incarcerated residents in mind?

Any prison in space, regardless of the architecture, will be extremely impractical. To explore the challenges posed by space prison, space lawyer Chris Newman proposed a hypothetical scenario in which he commits a crime in a space settlement: "What are you going to do to punish me? Are you going to lock me up? Well, that's a colossal waste of resources. I'm going to be sitting there, in a room, not cooperating, because I'm not going to be very happy that I've been found guilty . . . My role as mission specialist gets put to one side, and I then become a drain on resources. I become a machine designed to consume food and water."[9]

Newman identifies two key practical problems here. First, it takes resources to keep prisoners alive, resources that will be particularly dear in space. Not only does the community have to expend these resources for prisoner survival but it also needs physical space for the prison itself, since housing them in their former residences would likely be inefficient or dangerous. Living space will also be extremely scarce in an unterraformed space settlement, so the prison will need to be constructed by adding a pressurized module to the settlement or digging out additional space underground. This will require labor, another scarce resource in space.

If there are enough convicted residents, they could be directed to build the prison themselves, a scenario that is not unheard of on Earth. San Quentin State Prison in California was constructed in the mid-nineteenth century using materials quarried by the prisoners who were soon to be housed there and who were at the time living on the prison ship *Waban* off the coast of California.[10] Pitcairn Island, a tiny island in the South Pacific populated by descendants of mutineers from the HMS *Bounty*, had no need of a prison until the 1990s, when widespread accusations of child sexual assault emerged. The defendants in the related court cases constructed a prison while awaiting trial, and those convicted served their time within the very walls they had helped to build.[11]

Admittedly, the incarcerated residents of a space settlement would require food, water, air, and energy to stay alive whether they were in prison or not, but the larger challenge arises from the second problem that Newman raised in his hypothetical scenario: incarcerating the residents of a space settlement removes them from the settlement's labor pool. Depending on the number of prisoners and their levels of expertise, this could be devastating to the settlement.

Perhaps the inmates will be willing to continue working, despite their incarceration. Advocates for penal labor argue that it can support rehabilitation by teaching inmates skills that they can use once released, not to mention providing a distraction from other mischief they might get up to, like attempting escape or committing violence against fellow inmates. Cockell proposed that inmates in his extraterrestrial Exoconfac should all be "put to work," but only on tasks that do not facilitate escape or put the settlement at risk of sabotage, like "cleaning spare parts" or "sorting seeds for food growth."[12]

Advocates of penal labor also point out that it provides inmates with the dignity of meaningful work, a way to contribute to society even while they are removed from that society. Culturally, teamwork and self-sacrifice will likely be highly valued in space, where living on the edge of survival will make group safety more important than individual grievances. Prisoners in space may therefore be self-motivated to continue to work during their sentences, for the sake of the community. Even on Earth, without the hard vacuum of space waiting just outside the window to remind us of the importance of cooperation, incarcerated people still volunteer to work for the benefit of others. For example, many US states depend on incarcerated volunteers to help battle wildfires. Inmate firefighters perform dangerous, exhausting work during fire season, for as little as one dollar per hour, despite the existence of safer and sometimes better-paying jobs like working in the prison laundry.[13] They often don't even benefit from their training and experience after release; applicants with felony convictions are unable to obtain Emergency Medical Technician certification, disqualifying them from employment by many municipal fire departments.[14] So why do they do it? As inmate firefighter Marquet Jones explained in a *New York Times* article, "It feels good when you see kids with signs saying, 'Thank you for saving my house, thank you for saving my dog.' It feels good that you saved somebody's home, you know?"[15]

But what if the incarcerated residents of a space settlement refuse to work? There are plenty of ways the authorities could force an inmate to perform labor: by threatening them with a longer sentence, for example, or with the removal of luxuries or even necessities like food. Even if the settlement doesn't depend on the inmates' labor for survival, unproductive or "hard" labor has often been used by societies as a punishment in itself, useful for both deterrence and revenge. Forced labor has also been advocated as a tool for improving an inmate's moral character. This view is infamously summarized by the Nazi slogan "Arbeit Macht Frei" ("Work Makes You Free"), which was written over the entrance of several Nazi work camps, including Auschwitz. The irony of these words being displayed at concentration camps where prisoners were literally worked to death is obvious, but historian Otto Friedrich writes that the commandant of Auschwitz "seems not to have intended it as a mockery, nor even to have intended it literally . . . but rather as a kind of mystical declaration that self-sacrifice in the form of endless labor does in itself bring a kind of spiritual freedom."[16] The belief that hard work is character-building is not uncommon among people who aren't Nazis, but using this argument to force labor on incarcerated people "for their own good" is disingenuous, not to mention a common historical tool of oppression.

Mandatory prison labor is common in the US, but its proponents usually point to benefits like skills training and the satisfaction of meaningful work—similar to the arguments for voluntary penal labor—rather than "spiritual freedom." The state of Arizona, for example, requires all ablebodied inmates to work at least forty hours a week, and contracts that labor to a private company called Arizona Correctional Industries (ACI).[17] As recently as June 2021, the mission statement of ACI included creating "opportunities for offenders to develop marketable skills and good work habits," but also "achieving our revenue, profit and inmate work increase objectives."[18] (By February 2022, ACI's public mission statement had been changed to remove references to revenue and profit.)[19]

Profit is a key component of compulsory inmate labor. In the US, inmate labor is not covered by legal protections that guarantee safe working conditions, the right to organize, or a minimum wage.[20] In fact, the Thirteenth Amendment to the US Constitution, which banned slavery, specifically exempted unpaid labor "as a punishment for crime whereof the party have been duly convicted," so there is no constitutional requirement to pay inmate workers at all.[21]

Ratified at the end of the US Civil War, the Thirteenth Amendment left Southern states without access to the massive labor force of enslaved workers that their agricultural economies had been built upon, but it left open a window for an alternative solution: convict leasing. State legislatures began passing laws against vagrancy, and selective enforcement and discriminatory sentencing of these and other minor laws provided states with a population of incarcerated Black men who could legally be forced to work. Prisons leased these men to plantations, railroads, and mines at a profit, and the employers found the arrangement even cheaper than slavery, as they no longer needed to provide for the health or room and board of their workers. This form of forced labor differed from the earlier system of chattel slavery in that convicted laborers were not necessarily held in bondage for life, nor did their children inherit their enslavement. But in all other aspects, the comparisons were undeniable. As journalist Douglas Blackmon describes the system in *Slavery by Another Name*: "Free men, guilty of no crimes and entitled by law to freedom, were compelled to labor without compensation, were repeatedly bought and sold, and were forced to do the bidding of white masters through the regular application of extraordinary physical coercion."[22]

The practice of convict leasing was gradually eliminated in the early twentieth century, but the Thirteenth Amendment remains. Even when inmate workers in modern US prisons are compensated, it is often at a rate of as little as a few cents per hour. Defenders of these low wages argue that the state is already paying for the inmates' living expenses, so there is no need to pay a living wage. Opponents call this practice "modern slavery" and point out that inmates still require funds for court fees, items from the prison store like hygiene products, and support for their dependents on the outside. Nationwide prison strikes in the US in 2016 and 2018 cited unpaid inmate labor as a major grievance.[23] But this movement has struggled to make progress, given that state governments and private prison companies benefit financially from the practice. In California, the cheap labor of the inmate fire crews, who can account for 50 to 80 percent of the personnel on every wildfire, saves taxpayers around $100 million per year.[24]

Compulsory labor as punishment for a crime exists on a long spectrum, with Nazi work camps, Soviet gulags, and Chinese labor camps on one end, and community service sentences on the other. In between lie chain gangs, inmate firefighters, and convict leasing. Which of these practices, if any, will we take with us into space?

ALTERNATIVES TO THE CARCERAL SYSTEM

TRANSFORMATIVE JUSTICE

Aside from the practical problems of building and maintaining a space prison and dealing with the potential loss of the incarcerated residents' labor, another challenge of replicating our terrestrial carceral system in space is the harm that prisons themselves can cause in a society. "From the beginning of the sociology of the prison and research that's dedicated to understanding life inside prisons, we've seen [that] it's a dysfunctional environment," University of Tennessee sociologist and criminologist Michelle Brown said. "It's foundationally problematic. It severs social ties, it reduces opportunities for autonomy, various forms of decision-making. It's foundationally a space that's problematic."[25]

Prisons also tend to amplify inequalities that already exist in a society. "The disproportionalities that we face with regard to race, with regard to non-gender-conforming, trans communities, with regards to various axes of inequality . . . You tend to reproduce social divisions in these spaces [of mass incarceration] in ways that are heightened," Brown said. As civil rights lawyer Michelle Alexander describes in *The New Jim Crow*, Black Americans are arrested at higher rates and given harsher sentences than white Americans who commit the same crimes. Alexander argues that the resulting disproportionately Black population in prisons, combined with legalized discrimination against convicted felons who have completed their sentences, allows the US to maintain a Black "undercaste" without access to the ladder of socioeconomic advancement, all while avoiding legal challenges on the basis of racial discrimination. "Once you're labelled a felon," Alexander writes, "the old forms of discrimination—employment discrimination, housing discrimination, denial of the right to vote, denial of educational opportunity, denial of food stamps and other public benefits, and exclusion for jury service—are suddenly legal . . . We have not ended racial caste in America; we have simply redesigned it."[26]

But what is the alternative to prisons? One option is to banish the convicted resident from the space settlement. For example, they could be loaded into the next rocket and sent back to Earth. Transportation costs will likely make this prohibitively expensive, but on the other hand, kicking the resident out of the safety of the settlement without providing them with transportation to another habitable environment is essentially a death sentence.

Capital punishment is also an option, one that seems to be embraced by science fiction's darker visions of our future in space. In particular, disposing of enemies or political opponents via the convenience of the airlock is a staple of science fiction, including popular television series like *The Expanse* and the 2004 remake of *Battlestar Galactica*.[27] An earlier and frequently referenced example of murder-by-airlock is the classic science fiction short story "The Cold Equations" by Tom Godwin, in which a spaceship pilot discovers and then jettisons a young stowaway via the airlock, claiming that his actions are forced by circumstance—the ship's limited fuel and the "cold equations" of space travel physics. "It was a law not of men's choosing but made imperative by the circumstances of the space frontier," the narrator explains.[28]

"Frontier justice" in the mythologized American Old West, usually involving extrajudicial capital punishment, is often portrayed as an unfortunate necessity in the absence of a formal, "civilized" criminal justice system: the circumstances of the frontier—its remoteness and harsh conditions, its lack of official protectors—made these unofficial executions imperative, to maintain order at the edge of "civilization." It's easy to project this narrative forward onto future space settlers at the edge of human civilization, but it's also vital to remember that "frontier justice" was used to justify the lynching of Black men well into the twentieth century, long after the settling of the American frontier. As artist Ken Gonzales-Day notes in *Lynching in the West: 1850–1935*, "The irony is that while terms like 'frontier justice' or 'popular tribunal' still invoke images of cattle rustlers and stagecoach robbers meeting justice on a lawless frontier, they also mask a history of racial violence in a region that was not only culturally diverse but still is."[29] Will our descendants continue this legacy by condemning marginalized groups or political enemies to death in the name of the cold equations of space? The expediency of airlocks and the heartlessness of orbital mechanics will not absolve us of our responsibilities to our neighbors in a space settlement. If capital punishment accompanies us to space, it will be because we chose to carry it with us, not because it was forced upon us by physics.

Are these our only options? The "dysfunctional environment" of prison, or death-by-airlock? It can seem that way at first glance, because imprisonment or death have been the default punishments in most modern societies longer than any of us have been alive. Defenders of the carceral system consider the idea of a society without prison to be naïve, or as British politician Nick Herbert wrote, both "dotty" and "hopelessly utopian."[30] But this view merely reveals a lack of historical perspective. As Michelle Brown said, "The prison itself, as a mode of punishment, is relatively new in terms of our own history. It's something in the United States that we see as completely naturalized, but it's not been a primary factor across history and across various parts of ancient history." Brown studies the prison abolition movement, which advocates for the complete elimination of prisons. "It's a project that, in its current form, is like abolition around slavery, built around attempts to dismantle and disrupt current systems of power, but also to bring about what someone like sociologist W. E. B. Du Bois would call abolition democracy: bringing something new into being and new ways of being."[31]

I asked prison abolition advocate Walidah Imarisha to summarize her view on why we need to move beyond the carceral system. "I believe that prisons make us less safe, all of us," she said. "I think that there are many different ways that we can address harm that's done in our communities. And that we can heal communities and individuals when harm is done that do not rely on carceral or punishment mentalities like the prison system."[32] Imarisha is used to facing incredulity from people encountering the idea of prison abolition for the first time. One of the first concerns they tend to have involves safety: how can we keep our community safe if we can't lock criminals away? Imarisha explained that she usually offers a counterexample: "So many folks have experienced sexual assault. The vast majority of those [perpetrators] are never arrested, let alone convicted and end up going to prison. That's not what happens with sexual assault . . . We actually are already living with the realities that folks who are doing harm are in the same communities we are. And so I think that's an important framing, because it really breaks apart that connection that people make, that prisons are about our safety, which they are not."

As Imarisha explained, many communities already know how to manage behavior and relationships without the police or the carceral system, because a safe and fair formal criminal justice system has never been available to them. These alternative justice systems include approaches known as "restorative" or "transformative" justice. Rather than focusing on the relationship between the state and the offender in terms of either retribution or rehabilitation, restorative justice asks how those affected by a criminal act would like to see the harm they experienced be addressed and repaired. Transformative justice looks beyond individual relationships between victims and offenders to consider the social structures that cause harmful behavior and seeks non-carceral alternatives to preventing and ameliorating these harms.

Brown has observed that the needs expressed by victims and survivors of crimes rarely align with the punitive functions of the carceral system. "When you participate in these [transformative justice] trainings and workshops, the first thing you're going to be asked is, what does safety mean to you? And what we see uniformly, again, and nationally, most people don't say more police or more prisons . . . What we know victim-survivors uniformly want is not so much an arrest or even putting someone in jail. They want the violence to end, and they will opt for the option that promises that over a more punitive option in almost all the surveys and emergent research that we have."[33]

Restorative and transformative justice programs are already being tested and studied here on Earth. Research comparing restorative justice with conventional criminal justice in the US, UK, and Australia, found that restorative justice programs reduce recidivism rates for violent and property crimes as well as reducing PTSD in victims.[34] As we imagine the future of criminal justice in space, we should take the opportunity to study transformative justice programs on Earth, determine which methods work best in different environments and community sizes, and use these lessons to guide our designs for the legal structure of space settlements.

Given the relative youth of the carceral system, we can also look to alternative justice systems used throughout human history and in Indigenous and non-Western societies today. For example, Imarisha shared a story that represents the spirit of the transformative justice movement. This story was passed to her from an Eritrean friend, who described the wanza tree in small Eritrean villages as a gathering place for celebration and mourning:

And he said when someone had done something wrong, they would bring the person under the wanza tree, they would form a circle around that person, and then everyone in that circle would go around and remind that person when they were at their best. They would say, "Remember when I broke my leg, and you helped me bring my crops in?" "Do you remember when our crops failed, and you helped us survive the winter?" "Do you remember when . . . ?" And then they'd give the person the choice, "Do you want to be this person that we all know you are, who has been such an asset to our community and has helped us grow and thrive, or do you want to be this person that has done this harm? It is your choice. But we know you can do better."[35]

Imarisha was moved by this story, and the concept of reminding a person who has committed harm against a community that they have value and identity beyond that harm. "And I hope that this is the kind of ethos that folks can bring to any kind of transformative justice process," she said.

HOW WILL WE ADDRESS HARM IN SPACE?

VISIONS FOR THE FUTURE OF JUSTICE

The environments in which we build space settlements will be vastly different from those we've built on Earth, but human nature will be the same. Groups of humans will form communities, those communities will have certain norms of behavior (whether formalized into law or not), and they will occasionally need to respond to individuals deviating from those norms. In other words, space settlements will need some kind of system for dealing with residents who break the rules. Creating a justice system in space poses several practical and ethical challenges, especially if we attempt to simply reproduce a terrestrial prison system in the stark environment of space. On the other hand, given that there are no police or prisons in space today, we have an intriguing opportunity to decide whether we want to recreate these dysfunctional structures in our new societies, or whether this is our chance to try something better.

We have to do more than determine which systems we *don't* want to carry into space. As Imarisha said, "The concept of abolition has to be complicated. It's not just about tearing down but it's actually about building."[36] What kind of justice system do we want our descendants in space to experience? Does this answer change if we imagine them as the victims, the judges, or the perpetrators of crime? As always, we must consider the implications of the failure state of any proposed system. What happens to the falsely accused? What if the justice system is bent to the will of a tyrannical government and used to punish political enemies or minority groups?

Imarisha also pointed out the importance of asking these questions before we build communities in space. "Oftentimes we wait until there's a crisis to have conversations," she said. "When folks are at their most vulnerable, when folks are at their most scared, at their most defensive, that's a terrible time to have a conversation about what justice looks like in your community. So, these conversations have to start now." Fortunately, these conversations do not have to start from scratch. Societies in space will be new, but they will not be built in a vacuum (puns about the vacuum of space, aside). The carceral system is not the only form of justice that exists on Earth today, and certainly not the only form of justice practiced by humans throughout history. We also have entire fields of study working to understand what justice means and how it works best.

Criminologist Michelle Brown agrees that we should take advantage of the wealth of history and research we've amassed regarding human behavior and justice. We should "root our work in those histories," Brown says, and work to avoid repeating the harm that criminal justice systems have inflicted on communities in the past and present. In fact, Brown recommends stepping back from the simple framing of the problem that I've presented in this chapter ("If someone commits a crime in space, how do we punish them?"). Instead, Brown says, we should "move away from constructs of crime and punishment—and as a criminologist, that's basically ending my career, and I'm happy to do that, a lot of us are." Brown suggests that we begin thinking about "What will flourishing life look like in restricted contexts? How do we do the most with what we have? How do we live the best, and what are the conditions for that? So, inverting the criminological questions that we live with about crime and punishment. What are the conditions that will best help us flourish in the context we may find ourselves in, and how can we collectively pursue that?"[37]

This is work that will improve our descendants' lives, regardless of what planet they live on. Throughout the history of our species, we have learned to survive in barely habitable environments by cooperating, but where there is collaboration, there is also the potential for conflict. Protecting each other from harm, without inflicting further harm, is a daunting task we have yet to perfect. But as we work toward justice in our communities on Earth today, we are strengthening the skills that will one day allow us to thrive in space.

9

WHO'S IN CHARGE?

The year is 1789 CE, and you are a sailor aboard a British ship in the Pacific Ocean. The atmosphere on the ship has worsened since you departed Tahiti with your cargo three weeks ago, leaving that comfortable paradise for another long, miserable journey. You are awoken one night by the sound of fighting; several crewmen have seized control of the ship and are planning to set the captain and his loyal followers adrift. Who do you stand with? Will you obey Navy rules, and be put to sea with your captain? Or will you stay on the HMS *Bounty* with the mutineers?

The year is 1794 CE, and you, along with a hundred thousand other enslaved workers in the French colony of Saint-Domingue, have finally freed yourselves from the brutal domination of your former masters. But the rich sugarcane fields are a tempting target for colonial powers, and English and Spanish troops are already on their way, sensing France's weakening hold. Who should you align yourselves with? Who will give you and your children the best chance for a life of freedom and peace? Will your people ever have the chance to control their own destiny?

The year is 2100 CE, and trouble is brewing in your space settlement. The water supply has started to run low. The farmers are arguing for a bigger share to keep up with food demands, but the power plant engineers insist that they can't cut back their usage without risking outages. Meanwhile, Earth has refused to increase shipments of ice mining equipment until the trade negotiations have been settled. Someone will have to make a decision to balance the population's physical needs, the demands of various factions, and the relationship with Earth. But who?

———————

MOTIVATION

IN SPACE, WE GET ALONG OR WE DIE

In late December 1973, trouble was brewing between the astronauts on board the Skylab space station and NASA's Mission Control on the ground. The three rookie astronauts who made up the Skylab 4 crew had been in space for about six weeks and had been working nonstop. Their sixteen-hour workdays were scheduled to the minute by NASA, but motion sickness and the difficultly of performing tasks in space had caused the astronauts to fall behind, with no time to rest.[1] What happened next is not entirely clear. News stories later reported that the Skylab crew staged a strike, turning off their radio and taking an unscheduled day off to rest and enjoy the view.[2] Jerry Carr, one of the Skylab 4 astronauts, described the incident differently in his biography, explaining that the crew decided to observe a scheduled day off rather than work through it, and that the communication disruption was due to a misconfiguration of the radio.[3] What is agreed upon in these retellings is that NASA and the Skylab crew finally discussed their concerns and agreed that the astronauts would be given more flexibility in their schedule and more uninterrupted rest breaks.

The Skylab incident was no "mutiny on the *Bounty*," but these events do hint at potential conflicts to come in space and reveal certain characteristics of the space environment that will inevitably shape these conflicts. The Skylab 4 astronauts worked for NASA and were ostensibly required to obey Mission Control, but while on board the station, they were effectively out of NASA's reach. NASA has no means of forcing astronauts in orbit to do anything, which leaves open the possibility of a strike or work stoppage. On the other hand, NASA does eventually have some control over astronauts once they return to Earth. None of the Skylab 4 astronauts ever flew in space again, although it's not clear whether this was punitive or as a consequence of NASA's decreased flight schedule over the following decade.

Humans are social animals, and we form hierarchies to organize our behavior. But figuring out who is in charge, and how they will control other humans' behavior, is a problem we've grappled with throughout history. This question of organization is sure to plague us in space, as well, and it will be vital for both the survival and the quality of life of space settlers. For example, the structure and actions of the government of a space settlement will be crucial for protecting the human rights of its citizens, including the rights I've been discussing throughout this book: governments will settle property disputes, regulate the economy, and determine the structure of the settlement's criminal justice system.

There are also countless small, practical reasons for developing a consistent set of rules in space. This won't be an environment where new arrivals can simply scatter across the surface of a planet or moon and do their own thing, living off the land independently; the only way to survive will be to pool resources. Settlers responsible for growing food will need to agree how to share water and soil nutrients to produce the most efficient yields. Someone will need to decide how to distribute the energy collected from solar panels or nuclear reactors to ensure that the population does not starve, suffocate, or freeze due to lack of power. In *The Consequential Frontier*, Peter Ward quotes space policy analyst James Vedda speculating about the tiny but vital decisions that will need to be made in a shared space community, like determining the temperature and atmospheric composition of the habitat, or regulating the design of docking collars on space ships and stations: "If they get into trouble up there, you can't have half a dozen different private stations up there and then each one has its own proprietary docking collar and they can't help each other if there's an incident."[4]

On the other end of the spectrum from basic practical needs, choosing and maintaining good systems of government in space could help prevent the violent deaths of untold numbers of humans in potential future wars. In other chapters, I've discussed the importance of considering the "failure state" of different parts of a space society. For example, the worst-case outcome of a poorly structured economy is widespread inequality. The failure state of an unethical labor system is massive labor exploitation. But in this chapter, the failure state of governance is war, and war in space has the potential to be deadlier than anything we've seen on Earth to date.

LOCAL GOVERNMENT AND THE POTENTIAL FOR TYRANNY

THINK COSMICALLY, ACT LOCALLY

Space settlements will need some kind of local government, even if they are technically managed by nations or corporations back on Earth. This is partly due to the physical limitation of the speed of light, which will make immediate communication with leaders on Earth impossible. But even with near-instantaneous communication, the Skylab 4 incident demonstrates that a community out of reach of a distant government can and will make its own decisions about how much it will comply with that government's orders.

Many space settlement enthusiasts are excited about space because they see it as a blank canvas where humans will be able to experiment with new and untested forms of government. But the first generation of settlers will not be a blank canvas; they'll build their local government based on the knowledge, experience, and culture they bring from Earth. As space scientist David Baker writes, "Designing an extraterrestrial society from scratch would be implausible, selective amnesia being a prerequisite for that. But the fact that lessons and models from the past on Earth could be brought into play would give advantage to the new communities."[5]

In addition, a ship that takes a group of settlers into space will operate under its own set of rules. For any settlement farther away than Earth's moon, this journey could take months to years, so by the time the travelers reach their destination, they will have been living in the governance structure of the ship for some time. Traditionally, ocean-going vessels on Earth tend to be run with a military-style hierarchy, even civilian ships. There is a captain at the head, and the crew follows a chain of command. This allows the crew to respond quickly and decisively in emergencies. Space crews, whether military or civilian, have followed this pattern for similar reasons, and even the International Space Station has an appointed commander at any given time. This tradition is likely to continue aboard ships ferrying passengers to space settlements. How will this governance system change as the group of settlers shifts from operating as a crew to identifying as a community of civilians?

Predicting or designing the government of a future space settlement from here on Earth today is a game enjoyed by both science fiction authors and political scientists. But ethicist Tony Milligan worries that this exercise can represent "a case of the wrong sort of science fiction. The sort that reproduces thinly-disguised versions of present day attitudes and then mistakes them for prophecy."[6] It's easy for a libertarian here on Earth to find themself arguing that a libertarian system is obviously the best option for surviving the space environment, for example. The same is true for proponents of any political philosophy. But we should not mistake these political preferences for prophecy by claiming that any particular system is inevitable. There is no way to exactly predict the values, priorities, or ideologies of future generations living in space, nor how the space environment itself will shape the needs of a community in ways that we have never experienced on Earth. Will the harsh conditions require a more equitable, cooperative society, or will they stimulate the rise of an authoritarian, dictatorial government? Will the vast freedom of movement available in all directions lead to more independence and liberal values, or will the restrictions imposed by the pressurized walls of the habitat enforce conformity instead? How will political structures shift if an emergency like a crop failure or disease outbreak strikes, and how will the system evolve over time as the population and scale of the settlement increase?

Not only is it impossible for us to predict exactly what space settlements will look like, and thus what kind of systems of governance will help them thrive, but it is also impossible for us to enforce any system we choose on our descendants. They will be the ones who will have to decide what works best for their survival and their ethics. But that doesn't mean we can't work to provide them with the greatest opportunity for freedom and happiness. Space ethicists Tony Milligan and James Schwartz both suggest viewing space through the lens of philosopher John Rawls's "veil of ignorance" to minimize our own biases.[7] In this model, previously mentioned in chapter 5, we imagine that we will eventually be living under the governance system we are designing but that we won't have any control over our place in that system. This exercise should lead to a fairer government, since we will want to minimize the unhappiness of the least powerful position in the community, given that we might end up in that position. However, even a system specifically designed to maximize values like fairness or liberty will likely evolve and change as the generations pass, so another important component of this design is to safeguard against potential corruption. In particular, we can try to anticipate potential sources of oppression as we design first-generation space habitats, economies, and legal systems.

The inhospitable environment of space may be fertile ground for tyranny. Astrobiologist Charles Cockell has frequently observed that the lack of abundant, naturally breathable air will make space settlements especially vulnerable to whoever controls the oxygen supply: "Required on a second-to-second timescale by all people, organisations that control the supply of oxygen have power over life with a thoroughness that few other commodities can command."[8] A rebellious population can live without food for several weeks, potentially giving them enough time to organize their own supply, but humans can only live for minutes without oxygen.

Dissent by oppressed populations is a powerful tool against tyranny, but only when it is allowed. It will be even easier in space to argue that certain forms of protest are too dangerous to be permitted, given the fragility of habitats and life support. This won't be merely a political excuse to suppress public dissention; protest actions like infrastructure sabotage and work stoppages could actually be deadly in a space settlement. Arguments about allowed forms of protest are not limited to dissent against the state; corporations and employers are also potential targets for these actions. In chapter 7, I argued for the importance of organizing for labor rights in space, but even a nonviolent strike could threaten the safety of an entire settlement's population if the workers are, say, water recycling technicians, or power plant engineers. Perhaps this will mean that striking workers will wield more power in negotiations, but it is just as likely to mean that their actions will be deeply unpopular with the population.

Cockell suggests that one solution is to provide and maintain "effective channels for dissent," such as a free press or a mechanism for open deliberation, to ensure that the population has a way to make their voices heard without threatening the safety of the settlement.[9] Similarly, Tony

Milligan argues that dissent in space settlements, which is vital to protect the liberty of future generations of residents, should be "tolerated and constrained," and that both the tolerance and the constraint should be built into the legal system, rather than simply hoping that cultural norms will keep everyone safe: "Tolerance for dissent or more specifically for *certain kinds* of dissent is something which is best structured into the political process in ways which do not leave goodwill (a scarce resource at the best of times) to do most of the heavy lifting," Milligan writes.[10]

CURRENT LAW GOVERNING SPACE

REVISITING THE OUTER SPACE TREATY

Today, we have no permanent human settlements in space and thus no local space governments. But that doesn't mean that human activities outside our atmosphere occur outside the law. While the Outer Space Treaty of 1967 forbids "national appropriation" of territory in space, it also requires that governments on Earth be responsible for their citizens' and corporations' behavior off Earth. Article VI states that nations that have ratified the treaty "shall bear international responsibility for national activities in outer space . . . whether such activities are carried on by governmental agencies or by non-governmental entities."[11] In other words, the US is responsible for the activities of a US-based company acting in space. Article VI goes on to say, "The activities of non-governmental entities in outer space . . . shall require authorization and continuing supervision by the appropriate State Party to the Treaty." "Authorization" is a relatively straightforward requirement; for example, companies launching spacecraft from the US today must first obtain permission from the Federal Aviation Administration. But "continuing supervision" is not a well-defined or agreed-upon term. Will a government have to send its own spacecraft to the Moon to monitor private mining operations there, or can it simply require regular reports from the mining company? Once settlement begins, will government agents need to be present to monitor the activities of the settlers?

No one yet agrees on the answer to these questions, which haven't been tested in the courts. "The question is whether [Article VI] requires a heavy regulatory hand, or nothing, which is my own view, in terms of what the private sector can do under current law," space lawyer Laura Montgomery said.[12] Montgomery worries that interpreting Article VI to mean that every single human action in space must be supervised by a government on Earth would be unnecessarily expensive and overreaching. As she noted during testimony before the US Congress in 2017, "Life is full of activities, from brushing one's teeth to playing a musical instrument, which take place now without either federal authorization or continuing federal supervision. Just because those activities take place in outer space does not mean they should suddenly require oversight."[13]

The Outer Space Treaty was written at a time when the major worry about the use of space was its potential as a tool in the Cold War, so its articles focus on relevant concerns like national appropriation of territory and the use of nuclear weapons. But today, the main driver for increased human activity in space is the potential for profit. And it's not just space companies that are trying to figure out how to make as much money as possible under existing law; governments are also turning to the stars with dollar signs in their eyes. In *The Consequential Frontier*, Peter Ward describes a 2017 US Congressional hearing held to discuss possible changes to the Outer Space Treaty to lessen the regulatory burden on the growing American space industry. Surprisingly, Ward notes, executives from space mining companies like Moon Express and Planetary Resources (now defunct) who attended the meeting argued for the treaty to remain unchanged. "These companies embrace the ambiguity of the Outer Space Treaty because it contains just enough room for more lax laws on a national level," Ward writes.[14] In recent years, the US government has made an effort to clear the legal path for space mining companies, passing the Commercial Space Launch Competitiveness Act in 2015, followed by the 2020 proposal of the Artemis Accords, an international legal framework that would allow companies to extract and use space resources. So far, however, other major space powers have been skeptical of these proposed laws, with the head of Russia's space agency referring to the Artemis Accords as an "invasion."[15]

Ward draws a parallel between the disparity between national laws regarding space resource extraction and the variety of maritime laws on Earth. The latter has led to the practice of "flags of convenience," in which shipping companies base their operations in certain countries (thus flying the flag of that country on their ships) to take advantage of their lax regulations. Ward speculates that this practice will become popular in space, as well, noting that "places such as Luxembourg are passing light regulation laws in order to attract space companies."[16]

Depending on how the phrase "continuing supervision" in Article VI of the Outer Space Treaty is interpreted in the future, private companies may have more freedom than governments in their activities in space. A space settlement on Mars cannot become the fifty-first state of the US, but it will still need some kind of local government, as the US itself will be six months' travel time and several minutes' communication time away. Perhaps the space companies responsible for founding the settlement will fill the void themselves. Ethicist Erik Persson has explored the idea of a corporate-run space government and noted that it "might in fact be very convenient. A company operating on another world already has command structures, administrative structures, personnel, infrastructure, etc. in place."[17] In fact, Persson observes, there are numerous historical parallels in which private companies were placed in charge of European colonies, such as the East India Company, the Hudson's Bay Company, and the British South Africa Company. However, Persson concludes that a corporate-run government in space is "not a good basis for civil liberties" for a number of reasons: First, the power in such a settlement would be concentrated in an organization that is both the sole employer and the main supplier of life support. That organization's top priority, by definition, would be profit rather than the civil liberties or safety of its residents. Also, the leaders of the settlement's government would be selected by the corporation's leadership, likely back on Earth, rather than chosen by the residents they govern.

SpaceX drew public attention to the question of space governance in late 2020 when it released the terms of service for its satellite internet service, Starlink. Keen-eyed readers quickly noticed a section that stated, "For Services provided on Mars, or in transit to Mars via Starship or other spacecraft, the parties recognize Mars as a free planet and that no Earth-based government has authority or sovereignty over Martian activities. Accordingly, Disputes will be settled through self-governing principles, established in good faith, at the time of Martian settlement."[18]

Antonino Salmeri, one of many space lawyers to respond to this claim of Martian independence, notes that the Outer Space Treaty, which states that international (Earth) law applies in space, "naturally take[s] precedence over contractual terms of services," thus rendering this clause invalid.[19] But Salmeri also observes that "any attempt to escape international law on Mars may actually turn out to be strategically counterproductive," since the UN Charter already states that an independent community of humans is entitled to political independence. While it takes time for this autonomy to evolve, terrestrial international law may one day be the very mechanism that recognizes the right of self-determination for a community of space settlers.

HISTORICAL COLONIES AND THEIR PARENT GOVERNMENTS

ARE INDEPENDENCE MOVEMENTS INEVITABLE?

To explore how the relationship between space settlements and their parent governments on Earth might evolve in the future, I asked Lauren Benton, a professor of history and law at Yale University, how colonies and their parent empires interacted in the past. "It's difficult to reach broad generalization about this, [but] I can tell you about some of the patterns," she said. "[Colonists] relied on a lot of the routines that they used in governance at home and constituted their settlements as extensions of the realm."[20] This is, roughly, the model of our current legal structure under the Outer Space Treaty: activities in space are supervised by the relevant governments, essentially acting as extensions of those nations (without the ability to claim national territory).

On the other hand, Benton pointed out, empires of the past did not have anything like the Outer Space Treaty to guide their actions. "None of the powers engaging in colonization actually had a very clear-cut constitutional or governance model for how they were going to do this. So really, they were shifting all the time. They would create a certain framework for governance in the colonies and then forces would come to light that would cause that framework to need to be adjusted." Benton offered the example of colonial elites gaining power and wealth, and then demanding greater autonomy from their parent government. In space, external forces are also likely to alter the framework of governance. The discovery of an unexpected deposit of minerals or other resources near a settlement could produce a sudden shift in the economic balance of power, perhaps prompting the settlement to demand more autonomy from governments or corporations on Earth.

Will these demands eventually lead to full-blown independence movements in space? This seems to be a common assumption among space enthusiasts and science fiction writers, likely a result of the influence of the US's cultural obsession with its own revolutionary origin story. English science fiction writer Stephen Baxter has observed that, "in American-dominated mid-twentieth-century SF [science fiction], 1776-style rebellions of near-future space colonies against the centre were represented as something of a default, a theme picked up by authors from a surprising array of backgrounds."[21]

This unexamined assumption that independence struggles are inevitable is also widespread in public perceptions of history, not just predictions for our future in space. "I actually think it's a little bit of an odd impression that people sometimes have," Benton commented. "Particularly Americans who are focused on the American Revolution tend to view this as the most central preoccupation of colonies." Benton explained that this is a misconception. "The colonies very often depended very much on trade, and their elites depended [on their home country] for social advancement and economic advancement." Rather than trying to pull away from their parent governments, these elites sought approval and favor from their distant rulers. "Elites really dreamt, very often, of returning home with their fortunes and sending their children back home to be educated, and marrying people back home, and trading back home."[22]

For the first generation of space settlers, in particular, ties to their home planet will likely remain similarly strong. These bonds will not only include political or economic ties but also cultural and familial relationships. Space lawyer Laura Montgomery noted that corporations operating in space will also be bound by their financial ties to Earth. "If [the corporation] is doing something that violates what the actual laws are, then its assets can be attached, it can be fined, by governments here on Earth. So I think it will be a long time before we see any kind of declarations of independence."[23]

This doesn't mean that there will be no conflict between settlements and their supervising governments on Earth, even in the early days when the relationships with Earth are strong. Benton offered the example of a phrase used often in the Spanish Empire, "Obedezco pero no cumplo," or "I obey but do not comply." Benton explained that this phrase represented "an interesting game that governing elites would play, where they would receive instructions from the home government . . . and they would sort of do what was asked of them, or pretend to do what was being asked of them, but not very effectively, not very quickly."[24] This obedient non-compliance is similar to organized labor actions like work slowdowns or "work-to-rule," in which workers perform only the bare minimum to meet the requirements of their contract, obeying the letter of the law but not the spirit.

Benton also described another tactic used in European colonies that would be particularly well-suited to space: "They were very far away, so they could always ask for clarification. So, they would write back and say, 'We're not really sure what you meant by that. Could you give a little more specificity here?' And they knew that it would be months for this message to travel back to Spain or to England or France and then months for the answer to come back. So just by that very delay, they were gaining a lot of time during which they could do what they wanted."[25] Communication delays between the Earth and nearby settlement sites like the Moon or Mars would be much less than months, but it seems likely that distant settlements might take full advantage of the limitations of lightspeed, asking for further clarification from Earth while they continue to make their own decisions.

Note that these particular historical arguments view the parent/colony political relationship from the point of view of the colonial elites. However, history has shown us that the oppressed or enslaved members of a colonial population, who benefit much less from their relationship with the parent nation, can also be drivers of revolution. For example, the Haitian Revolution, which eventually resulted in Haiti's independence from France, began as a slave revolt in 1791. Many of the enslaved Africans who rose up against their white French overseers in the colony of Saint-Domingue were fighting for French citizenship and equal rights under the law. But when Napoleon Bonaparte sent troops to Saint-Domingue a few years later to reestablish French control and, secretly, to reinstitute slavery, he triggered a bloody war that eventually led to France's retreat and the declaration of Haiti as an independent nation. While space settlements may have financial and cultural incentives to maintain a close relationship with their terrestrial nations, a settlement with an oppressive government or highly unequal social structure may produce a similar uprising.

Putting aside the question of revolution, Benton also suggested an alternate source of potential conflict in space. "I would mention another element of strife which I think might play itself out in the future in space colonization and was very present in the history of colonization, and that is the conflict among and between empires for claims to colonies and settlements . . . That was the main focus, I would say, of colonial governance right up until the end of the 18th century." This was, in fact, one of the major motivations for the Outer Space Treaty: the concern that the US and the Soviet Union would extend their Cold War into space. It's not much of a stretch to imagine space settlements being used as pawns in a conflict between their parent nations back on Earth.

WAR

IS WAR IN SPACE INEVITABLE?

Even more common in science fiction than the idea of revolution is the trope of violent, military conflict in space—a kind of "star war," if you will. How dangerous would a war in space be? The writers of the Outer Space Treaty were particularly concerned about the combination of space—with its ease of access to the surface of the Earth—and weapons of mass destruction. Not only did the threat of nuclear weapons orbiting overhead have the potential to worsen Cold War tensions but experiments with high-altitude nuclear explosions had created temporary radiation belts that had caused several satellites to fail.[26] It was clear that nuclear weapons had the ability to contaminate the space environment, damaging satellites and potentially limiting human spaceflight, as well as posing a risk to life on Earth.

But a combatant situated in space doesn't need a nuke to cause terrible destruction on Earth. Our planet sits at the bottom of a gravity well; it takes much, much less energy to deliver a chunk of mass to the surface of the Earth from space than to launch it into space from the surface. A space combatant doesn't even need bombs to damage the Earth's surface (and the cities and populations that exist there); they can simply drop rocks on it. Nature has already demonstrated that the kinetic energy imparted by a collision with a rock from space can cause widespread geological damage, catastrophic climate change, and extinctions. Humans, unfortunately, have also learned the power of repurposing large objects like commercial airliners into deadly and destructive kinetic weapons.

Planetary scientist Ian Crawford and science fiction author Stephen Baxter estimate that a ship designed to transport humans on a round trip from Mars to Earth (specifically, the VISTA ship proposed in 1998), would carry enough kinetic energy to destroy a city if it collided with the surface of a planet. In fact, the energy of such a collision would be roughly equivalent to ten times the energy released by the atomic bomb dropped on Hiroshima.[27] A space settlement wouldn't even need to waste a ship to deliver this level of destruction to Earth, given the amount of free-floating mass already available in the asteroid belt. In this hypothetical future, asteroid miners may already have developed more efficient ways to move giant rocks around in space for their own work, so the technology to divert an asteroid onto a collision course with Earth or another target would likely already exist.

Space settlements themselves, while not sharing Earth's vulnerable position at the bottom of a gravity well, will nonetheless also be extremely fragile environments. Whether they exist as space stations or on the solid surface of a planet or moon, these populations will already be fighting a defensive war against their deadly surroundings. An attacker wouldn't necessarily need to physically destroy a settlement to kill its population; they would only need to find a way to poke a hole and let the air out. Baxter has observed that "throughout history humans have shown a willingness to use the supporting environment as a weapon, even at the cost of wrecking it."[28] Will future space combatants mirror the leaders of our past, who burned crops, tools, and civilian infrastructure to incite famine and subjugate their enemies? In the event of a retreat or rebellion, will our off-world descendants practice a scorched-earth policy (scorched-Mars policy?), smashing life support equipment, depressurizing habitats, and exposing farmland to radiation?

Based on their calculations of the catastrophic kinetic energies involved in space travel and their implications for interplanetary warfare, Crawford and Baxter conclude that "human affairs in an extraterrestrial context cannot be conducted through warfare, which is more likely to destroy the contending cultures and perhaps extinguish mankind altogether than to lead to any desirable political outcome."[29] But disagreements in space will surely be impossible to avoid; there will be ideological differences, disputes over limited territory or resources, and economic conflicts. We've had thousands of years of war on Earth; isn't war in space inevitable?

"People tend to think that we're always going to have war, even war in outer space," anthropologist Brian Ferguson said. "They think it's in our nature, in part because they think that humans have always made war . . . I can tell you quite definitely that this is not the case."[30] Ferguson studies the anthropology of human warfare and explained that this misconception is unsupported by our own history. "For instance, I have written about an area called the Southern Levant," Ferguson said. The Southern Levant is located roughly in modern-day Israel, Palestine, and Jordan. "And the incredible thing, I believe, is that from 13,000 BC to 3,200 BC, there is not one single place where you can say that war was present . . . and in other parts of the world [in this era], definite evidence is abundant. And what I think you can discern there is that those people, the people who domesticated plants and animals, also found ways to domesticate conflict."

Ferguson acknowledged that conflict itself is likely inevitable, and that humans obviously have the proven ability to go to war. "But what I don't think you can argue, if you look at the scientific evidence, is that we have some internal disposition to prefer solving problems by killing members of other groups. And if you don't have that, then you can imagine a future without war." While we are imagining our future in space, with technologies, lifestyles, and cultures that don't yet exist, perhaps it's not too optimistic to imagine that future without the ever-present threat of war. In fact, imagining and planning for a war-free future in space may be crucial to the survival of our species, given the enormous potential for destruction and death.

DEFINING OUR VALUES

For many space settlement advocates, one of the major benefits of founding new societies in space is the opportunity to experiment with different—or even brand new—forms of government. Human history is essentially a laboratory for these experiments, but in modern times, attempting a new form of government requires getting rid of the old one, a process that is often bloody. As Erik Persson notes, "Having access to a completely 'new' world [in space] where one could construct the ultimate political system without having to overthrow any existing power structures would provide a unique opportunity to try out new ideas in a more peaceful way."[31] On the other hand, despite the vast amount of effort, thought, and argument that has been spent on speculating about the best government to use in space, today's futurists cannot choose our descendants' political systems for them. This is likely for the best; here on Earth today, most of us are unable to even predict the results of the next election, let alone the political circumstances in which humanity's first space settlements will be founded. It's even harder for us to imagine and understand what it will be like to live in space and what the culture and priorities of off-Earth populations will be.

However, we can work to minimize the chances that these future societies will succumb to tyranny, oppression, or war. For example, Charles Cockell and others have proposed habitat designs and technologies that could help prevent tyrannical forms of organized violence and oppression. Cockell believes that "maximising liberty can be physically engineered as well as incorporated into the more abstract ideas of social and political arrangements."[32] He suggests designing modular habitats, with independent oxygen sources under local control. This would prevent the state, or any other organization, from using centralized control of oxygen production to suppress rebellion, punish political enemies, or commit genocide. Redundancy in design and equipment, while more expensive, would safeguard against mechanical disasters as well as human ones.

On the other hand, Stephen Baxter worries that, "in a sense, to engineer for 'freedom' within a confined habitat is itself an oppressive act."[33] Instead, Baxter suggests that the best way to design a space settlement to maximize liberty is to ensure that travel is as easy as possible, so that dissenters and oppressed populations can escape tyranny. Libertarian futurist Paul Rosenberg has made a similar argument, supported by historical parallels on Earth. For example, Rosenberg notes that ancient Egyptian civilization was geographically constrained to the narrow region of fertile land surrounding the Nile River, while ancient Greece was spread across the shores of the Mediterranean Sea. He argues that these environmental distinctions were partly responsible for the differences in culture between Egypt and Greece: Egyptians, who were tied to their land with little opportunity for movement, created a "static" society with a monolithic religion and a highly stratified social structure, while Greeks, who could travel much more easily between city-states by boat, formed a "massively decentralized culture" in which knowledge could not be tightly controlled by institutions. Based on this interpretation of history, Rosenberg argues that freedom of travel should be carefully protected in a space-based society, so that dissenters in one community can choose to leave to pursue a life more aligned with their values, allowing experimentation and innovation to flourish.[34] Cockell has suggested that freedom of movement can be protected in space through the design of the technologies that allow individuals to travel freely. For example, spacesuits should be mass-produced and easily maintained, comfortable enough to wear for long periods, and easily donned and doffed, thus lowering the physical barriers to free movement between habitats.[35]

Aside from technological solutions, we can also work to protect human rights in space through our legal system on Earth. Historian Lauren Benton pointed out that one advantage we have today compared to the age of colonization and imperialism in the past is our system of international law. "We did not have something recognized as international law in the early modern world. We had certain conventions, we had certain examples in Roman law and canon law for transpolity legal arrangements, and there was a practice of people trading across lines of culture and religion, but we didn't have anything recognized formally as international law. We do now."[36] As we continue to develop international law regarding space, we will be influencing not only human actions in space today but also the future of life in space as the law continues to evolve.

Space lawyer Chris Newman suggested that systems of law and government in new space settlements should be created carefully and deliberately, based on the values of the founders. "You need an enduring constitutional approach. You need a constitution and a basis of laws that has buy-in and acceptance from a broad amount of the population . . . People arriving into that colony or people being born into the colony will be inculcated with these values and be infused with the values that the basic constitutional documents hold."[37] We can't tell our descendants, several generations in the future, exactly how they should run their societies. But our actions, laws, and treaties today will influence the value systems of the founders of the first space settlements, and those values will be passed down to future generations via their culture and their legal system.

Inevitably, however, conflict is sure to arise in space once humans are living there permanently. Is there anything we can do now to help prevent that conflict from escalating to a potentially catastrophic war? "If we want to keep war from space, we have to defeat war on Earth," anthropologist Brian Ferguson said. "And while we can't see that happening now, that has to be the objective in the decades and centuries to come."[38] One of the concrete steps that Ferguson suggested to decrease the likelihood of future war both in space and on Earth is "participating in joint endeavors where common humanity is recognized." Ferguson offered the example of "the beginning of the United States' thaw with China [which] began with a ping-pong match." Space exploration has itself served as a joint endeavor between nations that recognizes our common humanity and our shared place in the universe. The space race that began between the US and the Soviet Union as a proxy for Cold War tensions has evolved into an understanding of space as "the province of all mankind," embodied today in the International Space Station, where astronauts from ten countries, including the US and Russia, have cooperated to build the station and conduct scientific research.[39]

One of the most important steps for preventing war, however, is to consciously reject violent conflict in our shared value systems. Culturally, humans celebrate many aspects of war, especially in the US, where we have generally not had to suffer the worst effects of war in our own territory. To prevent war in space, we must change that culture, which won't be easy. "No one's going to wave a magic wand," Ferguson said. "The struggle is something that has to be waged over the long term by lots of people and lots of small-scale interactions who can visualize the possibility—which right now, a lot of people can't—of a future without war." Despite the effort needed to make this cultural shift, Ferguson emphasized the importance of believing that this future is possible, which he sees as not a naïve utopian vision, but rather one based on our own history. "We had a past without war. It's not in our genes to make war. So we can envision a future [without war]. And how we could get there, I don't know how, and I don't know when. But the possibility exists. And that, I think, is something that we need to encourage people to recognize."

IV

HOW CAN WE LIVE WELL?

10

WHAT IF I WANT TO HAVE KIDS?

The year is 1965 CE, and you are a young farmer in rural India. You've just gotten married and are making plans to start your family, but some American visitors in your village have been handing out devices that they say will stop women from having children. The visitors have explained that there are too many babies being born in India too quickly, and that it is your responsibility to help by having smaller families. They don't seem to understand that with fewer children, you will not be able to work your farm to produce enough food. And without children, who will care for you and your spouse when you are too old to work, yourselves?

The year is 2013 CE, and you and your spouse have just been given some difficult news. Prenatal testing has revealed that your baby has inherited your genetic disorder, a condition that's left you with physical disabilities and some chronic pain. It's early enough that you can still legally terminate the pregnancy, if you choose, but you're not sure what to do. On the one hand, you would do anything to spare your child the pain that you live with. On the other, if your own parents had decided that your life would be too difficult, you might never have been born. How can you decide what's right for you, your spouse, and this fragile potential life in your hands?

The year is 2100 CE, and humanity's first space settlement has been running smoothly for a few years. The community struggled in its early days, but now the food supply is stable, and the life support equipment is robust and dependable. People are starting to look beyond the next harvest and make long-term plans, plans that include the next generation. But can the settlement spare the time and resources necessary to support a batch of hungry babies? Is it even possible to safely conceive and raise children in space? Who will be the first to take the risk?

———————

MOTIVATION

THE COMPLICATED ETHICS OF THE BIRDS AND THE BEES

A self-sufficient human settlement must, by definition, be able to sustain a constant population level. If the settlement can only maintain a steady population through immigration from Earth, it will never be able to stand on its own, independent of its planet of origin. Fortunately, the human body has a built-in mechanism for increasing population via reproduction, one that works so well that people frequently initiate the process unintentionally here on Earth.

But despite its relative ease as a process, human reproduction has been surrounded by a flurry of ethical debates for most of modern history. Cultural and religious taboos against extramarital, interracial, or interreligious procreation have burdened countless children with undeserved social stigma. The controversy over abortion rights in the US has motivated deadly terrorist attacks. Every new reproductive technology, from contraception to IVF to genetic screening, has introduced new ethical questions about bodily autonomy and the nature of parenthood. The growing understanding that we all must share this planet and its limited resources has motivated international attempts at global population control, while fears about the genetic heritage of our species have propelled eugenicists to commit atrocities. These issues can evoke strong emotions, to the point of violent conflict, because they exist in the tension between two fundamental, but often conflicting, principles: that a human has the right to determine what happens to their own body, and that we have a duty to protect the future of our species.

This key question in reproductive ethics—How do we balance the needs of the community with the rights of the individual?—will only become more crucial in a space settlement, whose health and safety will be precarious at the best of times.

CAN HUMANS REPRODUCE IN SPACE?

HOW DO WE EVEN ANSWER THIS QUESTION ETHICALLY?

It's not clear yet whether natural reproduction in space will even be possible for humans. Humans evolved in an environment with a constant gravitational pull of 9.8 meters per second squared, shielded from harmful solar and cosmic radiation by the Earth's magnetic field. Decades of medical research on astronauts living and working in space have revealed diverse and unexpected effects of the non-terrestrial environment on the human body.

"Every bodily system [in an astronaut], whether you're talking about their vision, or their nervous systems, or their immune systems, they're affected [by the space environment], and often adversely so," physician and space reproduction researcher Shawna Pandya said. "So, when you [ask] the question: Can we safely reproduce in space? There are so many unknowns. And you can't help but suspect that those effects that we see in fully developed adults would be amplified in developing bodies."[1]

None of the individual steps required for successful reproduction (copulation, conception, fetal development, birth) has even been tested on humans in space, let alone the process as a whole. Many of these steps are vulnerable to disruption by adverse environmental conditions or the physical health of one or both parents. The extra radiation in space might damage the parents' reproductive organs and decrease their fertility. Microgravity could make copulation difficult or interfere with conception or implantation. Development in a low-gravity, high-radiation environment could damage or kill the growing fetus. Bone and muscle growth during infancy and childhood could be inhibited in the absence of Earth's gravity, and children reared in space might suffer from an increased rate of cancers.

Many of these potential barriers to reproduction in space have been observed in animal studies: rats flown into space had decreased sperm counts, mouse embryos fertilized in simulated microgravity and then transferred to mice in normal Earth gravity resulted in lower birth rates, and the fetuses of pregnant rats flown in space developed deficiencies in their vestibular system, which controls balance and stabilization.[2] On the other hand, frozen mouse sperm flown in space later produced healthy baby mice, sea urchin eggs fertilized in microgravity later developed normally back on Earth, and Japanese Medaka fish successfully mated and produced healthy offspring entirely in space.[3] The evidence for or against the possibility of healthy human reproduction in space is far from clear.

We could begin to perform this kind of incremental study on humans without jumping headfirst into attempts at reproduction. For example, we could study the reproductive health of astronauts as they travel to and return from space (although privacy concerns have limited such research to date). Physician James Nodler has even suggested flying fertilized embryos that are already known to be nonviable to the International Space Station to observe the effects of the space environment on their development.[4] At some point, however, the only way to know whether we can conceive and develop viable fetuses in space is by trying it. But these open questions in space reproductive science are intertwined with and hampered by medical research ethics: How can we determine whether human reproduction in space is possible without risking harm to fetuses and infants?

The ethics of including pregnant volunteers in medical research trials is complex. On the one hand, pregnant research subjects and their fetuses are especially vulnerable to the potential harms related to research trials, and the fetuses and newborns affected by the research cannot consent to the risk they are taking on. On the other hand, excluding pregnant subjects from trials results in a dangerous lack of data about the effects of new medications or procedures on pregnant patients. By sheltering pregnant volunteers from the risks of a research trial, we also prevent them from experiencing the benefits of the research.[5]

"If you look at our research ethics principles and protocols on Earth, if anything, we're overly stringent when it comes to experimenting on fetuses and embryos and pregnant women," philosopher Quill Kukla said. Kukla studies bioethics at Georgetown University. "We let pregnant women engage in almost no research protocols. Even just completely theoretical, tiny risks rule women out from participating in research right now. And I think that's actually a problem in the other direction." But if we encouraged astronauts to attempt reproduction in space, Kukla said, "What we would, in effect, be doing is swinging all the way to the other side. It wouldn't be part of an official research protocol, but we would be doing incredibly risky research in this kind of uncontrolled condition. And that does go against what we think of as standard principles of research ethics."[6]

One or two case studies of pregnant space travelers would give us an initial glimpse into whether human reproduction in space is even possible, but a future space settler who wants to make an informed decision about whether to have a child will need to know the actual risk levels for the parents, fetus, and eventual infant. This will require population-level studies, involving much larger numbers of space residents attempting, and potentially failing, to conceive and carry a fetus to term. It seems likely that prospective parents in the early days of space settlement will have no choice but to take a chance on a potentially dangerous pregnancy before we have enough data to quantify the risk.

THE PRESSURES OF PREGNANCY IN A SPACE SETTLEMENT

EUGENICS REARS ITS UGLY HEAD AGAIN

Even if it does turn out that human reproduction is possible in space, pregnancy will probably be discouraged in the early days of space settlement, given the strain that the parent's pregnancy and, later, the needs of the newborn will place on the community. As Shawna Pandya described it, "Not only do you have a crew member who's likely going to have to go on some sort of maternity leave, [but] the rest of the crew is picking up the slack now and filling in for the duties of the crew member who's on medical leave. And then once the baby's born, there's now an extra mouth to feed."[7]

Eventually, the child will grow old enough to contribute labor back to the community, rewarding the investment of time and effort spent on raising them. But this kind of framing—the idea that a child's value lies only in the work that they may one day perform—can place enormous demands on an infant (and their parents) before the child is even born. For example, what will happen if a baby is born with extra needs in terms of care or resources? Will the settlers be willing to provide these additional resources, even if the baby's condition means that this societal "debt" will never be fully repaid?

"Presumably, at least at the beginning, there would be enormous and understandable pressures to produce healthy children who would be able to fully participate in continuing to found that colony," Quill Kukla noted. They pointed out that an early space settlement will have minimal infrastructure in place for medical care, social services, and educational services. "One of the concerns I have is [that] there's already, here on Earth, increasing pressure on prospective parents to abort fetuses who prenatal tests show [will] be disabled or have special needs. And I worry that there would be heightened pressure to abort disabled fetuses in space colonies, especially at the beginning, for resource issues, if nothing else."[8]

On Earth today, especially in richer countries like the US, our societies have enough excess resources to support members who are unable to contribute enough labor to balance their consumption. We don't condone the killing or neglect of children or adults who are considered unable to contribute, at least not legally. Murders of disabled people by their caregivers are unfortunately all too common in the US, and there is often a certain ableist public sympathy for the perpetrators, but such crimes are usually blamed on the caregiver's exhaustion or their belief that death is better than a life with disabilities, rather than their victims' purported debt to society.[9]

Despite the ableism still present in modern society, we generally recognize that every human life has intrinsic value, and that a person's disability does not counteract their right to exist. However, this moral calculus gets murkier when considering a *potential* human life rather than an existing one. In some countries, the birth of disabled children can legally be prevented after prenatal testing reveals a potential disability. One of the most common examples is Down syndrome, or trisomy 21, a genetic disorder that can now be detected during pregnancy. People with Down syndrome usually experience mild to moderate intellectual disabilities, distinctive facial features, and various physical disabilities. The life expectancy for a person with Down syndrome with access to adequate health care is over 60 years, but most such adults cannot live independently.[10] In

European countries where abortion is legal and prenatal screening widespread, abortion is very common after a diagnosis of Down syndrome. For example, in Iceland, 80–85 percent of parents consent to prenatal screening for Down syndrome, and nearly 100 percent of those diagnosed with the condition choose to terminate the pregnancy.[11] In the US, this issue is more controversial; only about two-thirds of parents who receive a prenatal diagnosis of Down syndrome choose to abort, and several states have attempted to pass selective abortion laws that would ban abortions due to diagnosed fetal disabilities.[12]

Despite the controversy surrounding prenatal screenings and abortion, it's clear that parents who are given the opportunity will often choose to terminate fetuses with diagnosed disabilities, even in countries that have the resources and capacity to support disabled citizens, due to the parents' concerns about the difficulty of raising a disabled child and adult. Will this pattern persist or even increase in a community living at the brink of failure in the harsh environment of space? Kukla worries that "this would enforce a horrific kind of ableism where we dramatically disvalue children who have special needs or who can't contribute in the same way to the propagation or founding of this colony."[13]

It's not too big a step from pressuring parents to abort fetuses with detectable disorders to screening prospective parents for the ideal genetics necessary for producing healthy, and more importantly, productive offspring. "It would be a breeding ground, so to speak, for the worst kind of eugenic impulses," Kukla said, "where people would start being under enormous imperatives to design specific sorts of babies that are going to have the right qualities to propagate in the way that's seen as valuable in that community. We have a long history of ethical concerns about that kind of eugenic program."

In chapter 3, I argued against the use of genetic screening for selecting potential space settlers. But even if a settlement's founders choose to establish their community on anti-eugenics principles, those values may evolve over time in response to the settlement's environment. Imagine, for example, that a disaster suddenly kills a significant portion of the population, threatening the diversity of the settlement's genetic pool. In order to regain a healthy genetic diversity in future generations, the settlement may be tempted to dictate which parents can reproduce, and with whom.

POPULATION CONTROL: PREVENTING OVERPOPULATION

TOO MANY HUNGRY MOUTHS TO FEED

Any newborn resident of a space settlement, able-bodied or not, will increase the demand on the community's shared resources. If the population grows too quickly, these resources may be consumed faster than they can be produced, leading to shortages and potentially threatening the health of the entire settlement. On Earth, overpopulation in a society is tied to an increased risk of famine and starvation, because throughout history, food has almost always been our limiting resource.

In space, however, drinkable water and even breathable air will also be precious and limited resources. A rapid rise in the birthrate will represent not only less food on everyone's dinner plate but also increased competition for oxygen. Human bodies can withstand long periods of famine, allowing us to survive the lean times until the next harvest or hunt. This provides a small buffer if variations in population level or food supply produce a temporary food shortage. But the window for surviving an oxygen shortage is much smaller and less forgiving. Space settlements will need reliable methods for keeping their birth rate low enough to avoid the deadly risk of overpopulation.

In the decades following World War II, international organizations attempted to address the population growth rate in the more fertile regions of the world with a series of escalating methods. Family planning advocates raised funds to provide low-income residents of developing countries with free educational materials and birth control equipment—such as condoms and, later, oral contraceptives—with the expectation that parents would choose to limit the size of their families if given the means to do so. But as the number of people on the planet continued to rise, some population control advocates pushed for more drastic methods. When free birth control and voluntary sterilization programs were seen as too slow, several countries implemented financial incentives for sterilization. Poor residents of these countries were offered an impossible choice between accepting a lifeline out of their desperate poverty and permanently surrendering their chance at future children.[14]

In some of the most populated nations of the world, governments decided that involuntary sterilizations were the only solution. During China's notorious one-child policy in the 1980s, for example, Shanxi Province instituted regulations stating clearly that the birth of a third child should not be allowed under any circumstances. Government policy required that "all women with one child were to be inserted with a stainless-steel, tamper-resistant IUD, all parents with two or more children were to be sterilized, and all unauthorized pregnancies aborted."[15] In India, compulsory sterilizations of men reached a peak during the national Emergency of 1975–1977, when Prime Minister Indira Gandhi suspended a number of civil liberties. Gandhi's son orchestrated a nationwide mass sterilization campaign, including vasectomy camps and strict regional targets.[16]

The population control movement of the twentieth century developed in parallel with many modern birth control technologies, including oral contraceptives and IUDs. In the space settlements inhabited by our descendants, these technologies will be available from the start. They'll also benefit from our modern understanding of demographic analysis. A settlement will be able to predict how large of a population it could support with its current and anticipated future resource supply and determine a maximum allowed birth rate to prevent overpopulation (with some wiggle room built in for uncertainties). Settlers could be encouraged or required to limit their number of children, assisted by freely available contraception and sterilization.

But in the hazardous environment of space, at the edge of human civilization, there will always be the potential for an unforeseen disaster like a habitat depressurization or a water reclamation system failure that would drastically reduce the available resources, and thus the number of settlers that the community could support. In addition, there may be settlers who refuse to comply with societal pressures to limit their number of children, as there always have been on Earth. What can a settlement do if its residents begin reproducing too frequently, despite the available contraceptives and guidelines?

Space settlements could choose to formalize their population regulation into laws, with consequences for parents who choose to have more than the recommended number of children. This would introduce a number of ethical problems. Laws regulating reproductive choices interfere with an individual's right to decide what to do with their own body. The controversy over abortion laws in the US today revolves around the question of whether a pregnant person should be forced to carry an unwanted pregnancy to term, and whether the fetus's right to be born overrules the parent's right to determine what happens inside their own body. In a space settlement, the dominant debate could instead center on the conflict between an individual's right to become and remain pregnant versus the settlement's duty to maintain a sustainable population level.

Another consequence of population-limiting reproduction laws is that the child at the center of the argument bears the brunt of the punishment. An unsanctioned child has no control over their own conception and birth, and yet they are born as the physical evidence of a crime. Depending on the society's consequences for breaking the law, such a child could be taken away from one or both parents or could be deprived of governmental financial support available to residents with fewer siblings. For example, in the 1960s and '70s, countries like India, Indonesia, and Singapore implemented policies denying government assistance in the form of maternity, housing, or tax benefits to state employees or citizens with three or more children.[17] Regardless of the precise legal consequences, the child will likely grow up with the constant reminder of the threat they pose to the community by simply existing.

But in a space settlement living on the knife's edge of failure and collapse, while punishing parents for bearing unallotted children may act as a deterrent to other prospective parents, by the time these children are born, it could be too late to prevent a dangerous strain on the life support system. Settlement leaders may be tempted to terminate unauthorized pregnancies before they come to term through compulsory abortions. "Whether or not you think that abortion is ethical under some circumstances," Quill Kukla said, "surely most of us agree that trying to pressure somebody into an abortion or force them into an abortion is always wrong. I'm concerned that people would be under huge pressure to do away with fetuses who were seen as unsupportable in these new environments where there weren't many resources."[18]

The history of international population control movements on Earth reveals another potential ethical threat. Many of these groups were motivated not only by concerns about the world's food supply but also by their fears of differential fertility: the observation that the population growth rates of some nations and ethnic groups were much higher than others.

Citizens of richer countries worried that the rapidly growing populations of poorer nations would recognize the strength in their own numbers and would pour across borders, demanding a larger share of the riches. Eugenicists warned that family planning education and contraceptives would limit the fertility rates of responsible, intelligent people without affecting the fertility rates of short-sighted, less genetically worthy parents, thus accelerating the deterioration of the human race.

Many population control organizations were not focused on reducing the global population growth rate so much as increasing the fertility of groups they deemed superior while decreasing the fertility of others. Reform eugenicists advocated for providing free day care and health care to encourage "fitter" parents to reproduce while simultaneously persuading "unfit" parents to have fewer children via family planning or even less voluntary methods.[19] The US passed the world's first eugenic sterilization law in 1907, and the Supreme Court later upheld the state's right to involuntarily sterilize people deemed unfit for parenthood, particularly people with intellectual disabilities. A wave of compulsory sterilizations followed the ruling, disproportionately affecting women of color, especially Black women.[20]

Could a similar system of forced abortions and sterilizations, perhaps motivated by eugenic beliefs, develop in space? Probably not initially, given that early space settlements will be small enough that a lack of genetic diversity may be a concern, and settlers might instead be encouraged to intermingle and reproduce across the population. But as settlements grow large enough for factions to form, with subgroupings based on religion, language, national origin, ethnicity, or some other characteristic, these factions may begin to compete for resources or political power and recognize the advantage of outnumbering their competitors. If we forget the hard-won lessons of our past, a new crop of eugenicists or population control advocates bent on demographic combat might emerge once again.

POPULATION CONTROL: PREVENTING UNDERPOPULATION

CONSENT IN SPACE

While overpopulation has been the primary concern of population control advocates here on Earth (eugenicists notwithstanding), space settlements may face the threat of its opposite: underpopulation. Even with advanced technology and robotics, human labor will be needed to repair equipment; care for children, the sick or injured, and the elderly; grow and prepare food; and perform the countless other tasks needed to maintain the settlement. An unbalanced age structure in the population—too many elderly residents with too few younger residents to care for them, for example—could produce a dangerous labor shortage. Underpopulation could also lead to a drop in the genetic diversity of the settlement, increasing the risk of genetic disease. And ultimately, a human space settlement cannot exist without humans: if a settlement's population dwindles too much, it will collapse.

As with overpopulation, underpopulation could be predicted, to some extent, by demographers in the settlement based on birth and death rates. But once the problem is identified, what approaches could the settlement take to increase its fertility? A first step might be decreasing the perceived burdens of parenthood; in today's economic system, examples would include offering free childcare and generous parental leave policies. More direct measures could reward parents for choosing to have children. Countries whose populations were severely depleted by World War I used a variety of incentives and disincentives to encourage reproduction. France outlawed not only abortion and contraceptives but even "anti-conception propaganda"—including birth control education—in 1920.[21] The Soviet Union banned abortion in 1936 and began offering financial incentives to encourage parents to have larger families.[22]

But suppose these measures fail to raise the birth rate high enough to sustain the population? Mandating reproduction would be an even more invasive policy than mandating birth control, and it would run afoul of similar issues involving bodily autonomy and consent. Advocates for abortion rights today argue that forcing someone to remain pregnant is unethical, although this position is not universal. On the other hand, nearly every legal system on Earth agrees that the process of forcing someone to *become* pregnant is immoral and illegal. As Quill Kukla explained, "There's a whole other cluster of issues around the fact that the way that we generally reproduce is through sex. So, if we're worried about pressure on people to reproduce, we also need to be worried about pressure on people to consent to sex in the first place."[23]

In chapter 3, I considered whether reproductive health and a willingness to conceive children would be a factor in crew selection for a new space settlement. The fear of underpopulation could make this a tempting choice. But biasing crew selection based on the potential number of children a candidate will have would more than likely produce other biases: against older candidates, disabled candidates, and candidates who are not interested in heterosexual reproduction. Aside from the ethical concerns about excluding certain categories of people from settling space, Kukla pointed out that limiting initial crew selection to only heterosexual candidates would not be foolproof. "People's sexuality is fluid," they said. "They may think that they're interested in heterosexual reproduction [initially]. Presumably, we're going to be mostly picking young, healthy people to go do this in the first place. And young people's opinions change, and their desires change. And they may get there and decide that heterosexuality is not for them after all." Reproductive health can also change suddenly and without warning, especially after exposure to the high-radiation environment of space.

More practically, it would be impossible to maintain this selectivity past the first generation of settlers, given that we can't screen for homosexuality or asexuality in the womb, nor would it be ethical to do so. But space settlers who are uninterested in heterosexuality may be faced with societal pressures to biologically reproduce, creating a new strain of homophobia in space settlements that struggle with underpopulation.

Perhaps space settlements will be able to address these particular ethical problems with advanced technology. The use of IVF and surrogacy, for example, could allow settlers to contribute to the genetic diversity of the settlement without having to engage in heterosexual reproduction directly. Science fiction imagines a future with artificial wombs, in which fetuses can be brought to term without inflicting either copulation or pregnancy on the child's genetic parents. But all of these technologies, real or imagined, require additional resources and advanced training. An underpopulated settlement dealing with a severe labor shortage may not be able to muster the personnel and equipment for these techniques and may fall back on the old-fashioned method for increasing the population. At that point, individual consent may be pitted against the survival of the settlement, and consent has historically been an early casualty in struggles for survival.

"Quite aside from the pressure to reproduce, I just worry about what would happen to our norms of consent under those conditions," Kukla said. "Already on Earth, it's the case that if people are in isolated conditions without a lot of oversight or without a lot of redress, there's a whole lot of sexual assault, there's a whole lot of sexual coercion. How effective would we be at enabling people to have real sexual autonomy in that kind of frontier condition when there isn't much set up yet, given how much of a problem it already is here on Earth?"

Female researchers in Antarctica, for example, have had to deal with sexual harassment and assault since they began visiting the male-dominated continent in the 1960s.[24] Sexual harassment is still extremely common in scientific fieldwork, where researchers travel together to remote locations around the globe with little protection from harassment by their supervisors or peers.[25] Space settlements will share many similar characteristics with fieldwork, including the isolation of a harsh environment and the potentially unequal power dynamics. Adding in a pressure to reproduce will likely increase the potential for sexual harassment, assault, and coercion, unless the settlement makes a deliberate effort to create a culture of consent and provide effective reporting mechanisms.

RAISING KIDS IN SPACE

THE RIGHTS OF THE NEXT GENERATION

One of the reasons that reproduction is so ethically complicated is that reproductive rights issues don't just involve the rights of the individual parents versus the needs of society—which is complicated enough as it is—but also the rights of the child at the center of the debate. Children have the same intrinsic value and rights that adults have, but they don't yet have the ability to consent to decisions that affect them. The adult members of a society have to figure out which choices are in the best interest of their children and hope for the best, always aware of the possibility that their descendants might grow up to resent them for their decisions.

Even though many children today would love to visit outer space, adults know that space exploration is currently too dangerous to risk their lives. To date, the youngest space traveler to reach Earth orbit was Soviet cosmonaut Gherman Titov, who became the second human to orbit the Earth at the ripe age of twenty-five, and eighteen-year-old Oliver Daemen became the youngest person to participate in a suborbital spaceflight in 2021. But if we intend to build communities in space, we will eventually have to subject children to the dangers of the space environment. We don't yet know how they'll respond: as with pregnancy, the only way to know for sure how children's bodies will adapt to and develop in space will be to try it.

"We [will] be producing new humans under conditions that we really don't understand," Kukla said. "And quite aside from what we can or can't ask women to do to try to control for risk, inevitably, we are going to be creating children under unknown conditions who might suffer pretty severe consequences from our not understanding the risk."[26] Aside from the physical risks, children in space might also experience stressful and potentially traumatic childhoods in early space communities fighting to survive.

Is it ethical to bring children into an environment with a higher risk of physical damage, mental trauma, and early death? This question is not unique to space. Throughout history, parents have moved their families away from the safety of their own childhood homes toward more dangerous environments—frontiers, cities, foreign lands—in the hopes that their children will have a better chance at success there despite the risks. Today, migrants and refugees must make the difficult choice between trying to raise their children in the midst of a war or climate crisis and risking a dangerous and potentially illegal border crossing to try to find a safer home. Young children have no say in these decisions that will affect the rest of their lives.

The special characteristics of the space environment do add a new wrinkle to this question, however. The bodies of adult astronauts who spend a significant amount of time in space change measurably in the absence of Earth's gravitational pull. Besides disorientation and motion sickness, space travelers can suffer from a loss of bone density and muscle mass.[27] Generally, these effects have been reversible, and astronauts gradually regain their normal bone and muscle health after returning to Earth. But it's possible that living in space for years could produce irreversible changes to the human body, and even more likely that a body that has never experienced Earth's gravity will never develop the strength to withstand it. In other words, children born in space may not be able to survive a visit to Earth. As Shawna Pandya described the issue: "If you give birth to a child in an altered gravity environment and somehow, they survive . . . could they ever come back to Earth and could they survive there? And if not, is it ethical to condemn another living being to not be able to come back to Earth?"[28]

The children of migrant parents on Earth are at least physically capable of returning to their original countries once they've grown up, even if political barriers remain. But the children of immigrants in space may not have the ability to survive a return to their parents' home planet. By choosing to bear and raise children in space, settlers may be depriving them of something we all take for granted: the ability to experience the environment we evolved in, to mingle with the planet-bound majority of the human race. Is Earth a birthright that we will be withholding from our descendants in space? Will they even care that they can never return to a world they've never seen? It's impossible to know, and we can't ask them. Even John Rawls's veil of ignorance is not useful in this scenario, as none of us can imagine exactly what it will be like to be raised off Earth.

HOW WILL WE PROTECT THE RIGHTS OF PARENTS AND CHILDREN?

POPULATION HEALTH BEYOND THE NUMBERS

In his book *Fatal Misconception: The Struggle to Control World Population*, historian Matthew Connelly notes, "When contemplating the seemingly inexorable rise in human numbers, the most thoughtful observers have eventually asked themselves: 'What are people for?'"[29] I asked a similar question in chapter 2: What are space settlements for? If they are meant merely as backups for Earth, to prevent extinction in the event of a planetary catastrophe, then the people who inhabit them could simply be considered biological storage mechanisms for the genetic heritage of the human race. But the human race is more than just our DNA. If we don't also pass down our culture, our values, and our ethics to future generations, what is the point? If the only way for human civilization to survive in space is by violating the reproductive rights of the parents living there, is it really worth the effort?

Instead of emphasizing the importance of maintaining healthy population numbers at all costs, future space settlements should consider how to protect the rights of the most vulnerable participants in the reproductive process: the first parents who attempt to become pregnant in space, providing invaluable data at great risk to themselves and their unborn children; settlers who want to make choices about their own bodies against the preferences of the community, including remaining childless or bringing children to term that the settlement would rather not support; the children born and raised in space, far from the natural physical environment of Earth; and particularly the disabled children who may face a culture of ableism insisting that they are burdens to the community. If the settlement makes a deliberate effort to emphasize that all human lives are intrinsically valuable, that all consensual reproductive choices are valid, this could lead to situations with an increased threat of dangerous labor or resource shortages. But at least no child will grow up believing they are unwanted by society and undeserving of existence. Isn't that worth the risk?

WHAT IF I GET SICK?

The year is 1793 CE, and you are exhausted. You've lost track of how long you've been working at a French aide station near the front lines of this never-ending war. There are always too many wounded men and never enough doctors. Before you lies an unconscious colonel with grave wounds. You know you could save him, but you'd need the help of at least three other surgeons, and they have their hands full with a dozen bleeding prisoners of war. Do you pull them away to save the French officer? Or let him slip away so you can save as many lives as possible, even though they are enemy lives?

The year is 2020 CE, and you are exhausted. The pandemic has reached your small, rural hospital, and your meager resources are already strained. No help is coming from the larger metropolitan hospitals, which are struggling with their own waves of virus patients. Before you lies a seventy-five-year-old grandmother who's been kept alive on a ventilator for three days. Without it, she would die, but the twenty-three-year-old man who's just been sent up to your ICU will also die without a ventilator, and you do not have one left for him. He has a greater chance of survival, and more life ahead of him, but can you bring yourself to end the woman's life support to save another?

The year is 2100 CE, and you are exhausted. The victims of the catastrophic decompression accident in Habitat 3 have been pouring in all day. One of your operating bays has finally opened up, and you're responsible for choosing the next patient to receive live-saving surgery. Two patients lie before you: one, a seventeen-year-old laborer, likely one of the poorer residents of Habitat 3, given his appearance; and the other, the commander of the habitat, a woman you've known and admired for twenty years. Without her, you don't know how the community will recover from this disaster. But surely the boy deserves a chance at life, too. Who should you choose?

————

MOTIVATION

THE SPACE TROLLEY PROBLEM

Many of the ethical problems described in this book are related to the scarcity of resources in a space settlement, such as living space, food, water, and even air. There is another category of resources that will be hard to come by in a space settlement but could mean the difference between life and death for its citizens: the resources related to medical care. These include supplies like antibiotics, blood products, or insulin; equipment like ventilators or dialysis machines; and trained personnel like surgeons or obstetricians.

On Earth, the availability of most of these resources depends on a long supply chain as well as manufacturing and training capabilities that won't be available in the early days of space settlement. Some supplies, equipment, and personnel will be brought from Earth, based on the expected needs of the settlement, but the demand for medical resources in space will not necessarily remain constant or predictable over time. A disaster like a fire, epidemic, or toxic chemical leak could leave a large number of residents in need of urgent care. Longer-term demands for medical care might also vary. For example, poor radiation shielding in a section of the habitat could cause a subsequent spike in cancer rates, or the incidence of a chronic condition like type 1 diabetes could unexpectedly increase among future generations in a settlement, ramping up the need for insulin. At some point, demand for medical resources in space will likely exceed supply, and settlements will need a way to choose how to distribute those resources ethically when lives are at stake.

The challenge of distributing medical care in an environment where resources are scarce often evokes the classic trolley problem in ethics. First described by philosopher Philippa Foot (using a tram instead of a trolley), the trolley problem is a thought experiment in which the driver of a runaway trolley realizes that the vehicle is barreling down the tracks toward five people who will all be killed if nothing is done.[1] The driver can't stop the trolley but they can flip a switch to divert the trolley onto a separate set of tracks, where only one person is standing in harm's way. The question posed by the problem is whether it's ethical to flip the switch, thus killing one person to save the other five. Utilitarian ethics, as well as many people's intuition, would hold that flipping the switch is the most ethical choice, as it would decrease the number of deaths in the scenario. But consider a parallel situation in medical ethics: Imagine that five patients are in desperate need of organ transplants. Would it be ethical to kill one healthy person to distribute their organs, thus saving the five sick patients in need of transplants?

The trolley problem and its variants are designed to probe our intuition about the ethical differences between harming someone by direct action or by inaction. Is flipping the switch to redirect the trolley toward a single victim more ethically justifiable than killing a healthy person to save five sick patients? What if the trolley scenario required the driver to push a single person onto the tracks to save five, rather than simply redirecting the trolley; does this change the morality of the action? These types of thought experiments also ask us to weigh the comparative value of lives, a common dilemma in triage and medical ethics. What if the single person on the trolley tracks was a loved one? What if they had a terminal disease that would kill them within six months? What if the five potential victims were children? What if they were convicted murderers? How can we decide which lives to save and which to sacrifice?

These are the choices that medical providers in scarce environments face on Earth today and will continue to face in future space settlements. Is there an ethical difference between allowing a patient to die by withholding lifesaving medication needed by someone else, and killing a patient by taking them off a ventilator in order to use it for another patient who is more likely to survive? If only one of two patients can be saved, does it matter if one is the leader of the settlement, the chief engineer, or the only surgeon, while the other is an unremarkable resident with no special expertise? What information will the medical team in a space settlement use to prioritize patients?

Like the trolley problem, these hypothetical scenarios about medical disasters in space settlements can seem farfetched and artificial. But even in the absence of a life-or-death dilemma, the decisions a settlement makes about how to distribute medical care will not only affect every member of the community but will also clearly demonstrate the settlement's beliefs about the relative value of its residents' lives.

PREPARING FOR MEDICAL EMERGENCIES IN SPACE

FITTING A HOSPITAL IN A SUITCASE

The job of ensuring adequate medical care for our descendants in space begins in the planning stages of a settlement back on Earth. Fortunately, we won't be starting from scratch, because we can study the lessons learned by medical providers who have treated patients in remote and isolated locations here on Earth. For example, the types of medication and equipment that should be included on an expedition to a remote location—on Earth or in space—depend on the size, environment, and anticipated length of the mission. A party of mountaineers climbing Everest needs to consider the possibility of frostbite and altitude sickness, while a group hiking through the Amazon will be more concerned with snake bites and tropical diseases. Long-distance space travelers will be more likely to worry about radiation exposure and decompression injuries. There are also many acute illnesses that can arise in any environment, like appendicitis or heart attacks. In the longer term, remote medical providers also need to manage chronic diseases like malnutrition, cancer, and natural aging. Since the founders of a space settlement will not be planning a return to Earth, they'll need enough supplies to see them through the early days of the settlement until they can either be resupplied from Earth or begin producing their own supplies locally.

Settlement planners will also need to consider medical expertise in their crew selection and training. As mentioned in chapter 3, cross-training will be crucial. If a settlement only has one trauma surgeon, for example, then more than one life may be lost if that surgeon is killed in a mass casualty disaster. A certain amount of cross-training can be done after crew selection; prospective space settlers can learn basic first aid while they study spacecraft operations and the basics of hydroponic farming.

But post-selection cross-training can only go so far, notes physician Ken Iserson, an expert in disaster medicine who has worked clinically or taught on every continent on Earth. "You can't cross-train someone to be a surgeon," Iserson said. "We do dental work [on expeditions], including extractions and fillings and other things. And that's cross-training. The physician generally does that. But you just can't train someone adequately to do general surgery in a week or whatever."[2] Settlement planners will need to consider medical training and experience during the selection process itself, rather than simply counting on mission training to provide enough expertise.

Redundancy in training also protects the medical providers themselves. After all, doctors are also humans vulnerable to disease and injury, and they might find themselves in need of trained colleagues. One of the most famous examples of this scenario in remote medicine is the case of Russian surgeon Leonid Rogozov, a member of the sixth Soviet Antarctic expedition. The only doctor on a team of twelve, Rogozov began suffering symptoms of appendicitis in April 1961, at the start of the Antarctic winter. With no chance of evacuation, Rogozov operated on himself, successfully removing the infected appendix with the help of a local anesthetic and the team's meteorologist and mechanic serving as assistants.[3] "I can't think of anything worse to do," Iserson said with a laugh while describing the incident. "Although, at one of our [Antarctic] stations a number of years ago, the physician actually came down with a large abscess in her mouth. She couldn't drain it herself. So, using mirrors and instructions, she had the other, non-physician clinician drain it."

Surgery—whether dental, trauma, or general—is not the only specialty that will be needed in a space settlement, where residents will live out their entire lives, encountering a variety of medical problems at different life stages. To provide equitable and sufficient medical care to all of its residents, a settlement will need expertise in obstetrics, pediatrics, and geriatrics, for example. Mental health care will be crucial as well, not least for the medical providers themselves, who will likely struggle with the stress of keeping the settlement's residents alive and healthy with limited resources.

Physician Jim Duff, an expert in wilderness and expedition medicine, said that while expedition doctors on Earth have to deal with unique and challenging medical situations in dangerous environments, family doctors face their own set of stressors: "They actually [have to] live with their mistakes," he explained. "They take a family and see them through various stages of life and then might end up palliating them in a terminal sense."[4] Doctors in a space settlement will have to deal with both: the unusual danger and scarcity of the space environment, combined with the responsibility of caring for their patients from birth to death. "So I would say it's going to be pretty intense in such a remote situation," Duff said.

Doctors in a space settlement may benefit from consultations with specialists back on Earth, although communication delays imposed by the speed of light will prevent this in the most urgent cases. Remote consultation and even telesurgery via robots are already used on Earth today to allow providers in rural or wilderness environments to take advantage of the knowledge and expertise of colleagues elsewhere on the planet. Iserson noted that one key ingredient for an optimal remote medical consultation is that the consultant should not only be an expert in their field of medicine but should also have a good understanding of the working conditions for the doctors in the field. "You want to have the best possible consultants available for the clinicians who are onsite," Iserson said. "And you want them to be as knowledgeable as possible about the situation at the South Pole or in space because otherwise, they'll be using a mindset from their civilian practices, which is totally inappropriate. And we saw that, certainly, in Antarctica."

Planning for a space settlement should include arrangements for consultations with physicians on Earth in the event of a medical case that exceeds the training of the settlement's clinicians. This kind of consultation has been used many times to solve engineering problems in space. The most famous case occurred during the Apollo 13 mission, when ground crews back on Earth helped the astronauts adapt their lunar module's carbon dioxide scrubbers after the crew were forced to use the module as a lifeboat following an equipment failure. Engineers on the ground had to help improvise a solution using only the equipment that the astronauts had on board. Similarly, medical consultants on Earth will need experience in the improvisational nature of remote medicine as well as a thorough understanding of exactly what equipment and supplies are available in the space settlement.

UTILITARIAN TRIAGE

BY THE NUMBERS

The simplest method for distributing a small number of indivisible benefits among a large number of people is through random selection. Lotteries are quick, transparent, and generally resistant to corruption or bias. Distributing scarce medical resources by drawing straws follows the principle, as described in John Harris's introductory medical ethics text, that "each person's desire to stay alive should be regarded as of the same importance and deserving the same respect as that of anyone else."[5] While not everyone can receive the same degree of medical care, they at least have an equal opportunity to receive that care. But equality is not equity; if some patients in the group are closer to death than others, offering them all the same shot at medical care is not justice and may result in more loss of life than necessary.

A first-come, first-served system, common in medical situations where scarcity is not an issue, like an ER on a slow night, has been called a "natural lottery," in which the randomness lies in the timing of each patient's medical emergency.[6] But in an environment of scarcity, a first-come, first-served system not only presents the same equity problems as a direct lottery but it is also biased in favor of any group with the ability to shove their way to the front of the line.

On Earth today, residents of richer countries usually only experience an obvious scarcity of medical resources during a mass-casualty incident like a natural disaster or act of violence. In such emergencies, the limiting factor for adequate medical care is the number of providers, who cannot treat all the patients before them at the same time and must decide which patients to transport or treat first. This can include a paramedic crew arriving on the scene of a mass shooting, for example, or an emergency room flooded with patients after a tornado passes through a residential neighborhood. Providers in these situations usually turn to traditional emergency triage, a method of prioritizing patients based on their needs and their chances of survival in order to maximize the number of lives saved.

The methods used in modern medical triage are usually traced back to Baron Dominique Jean Larrey, a French military surgeon during the Napoleonic Wars. The term "triage" derives from the French word "trier," meaning "to sort," and describes a system of sorting and ordering patients based on their needs. Larrey's battlefield triage rules stated that the most severely wounded patients should be treated first, and that those with milder wounds, who were less likely to die in the meantime, should wait for care. British naval surgeon John Wilson later argued in the mid-nineteenth century that patients who are likely to die even with medical care should also wait until the most immediate, but still salvageable, patients have been treated.[7]

These techniques, developed for military contexts, have been adapted for civilian use today. As Keith Abney, a philosopher at California Polytechnic State University, described modern triage: "[We] divide patients into groups and use what's called the 'rule of rescue' as the primary principle, such that we prioritize treatment of those for whom treatment is a life-or-death issue. And only after we've treated all of those folks, do we treat people with non-terminal, non-lethal problems, whether that's a broken leg or a cold. And then last, and least, we treat those who we cannot save their lives, whose condition is terminal, and who really have only palliative care."[8]

The goal of modern triage is to minimize the number of deaths, given limited personnel and resources. Medical science has refined and shaped the triage algorithms now used in various jurisdictions, each allowing providers to perform triage as quickly and objectively as possible. While the speed of triage ensures that treatment delays are minimized, the objectivity of the system has ethical benefits. Only the patient's current physical condition is considered by the provider performing triage; personal characteristics like ethnicity, occupation, and wealth are ignored. Larrey's original triage rules specifically stated that no regard should be given to the patient's rank or nationality during battlefield triage, a principle later formalized by the First Geneva Convention.[9]

This basic version of emergency triage can be expanded to consider not only the immediate survival of the patients but also their long-term likelihood of survival and health. Abney explained that in this method of triage, "We're going to give treatment out in such a way as to maximize the Quality-Adjusted Life Years of the patient. So if we only have enough resources for one cancer treatment, and there's a twenty-year-old [patient] and an eighty-year-old [patient], then the twenty-year-old gets it."

The quality-adjusted life year (QALY) is a concept that attempts to capture, in a single number, both the length of life and quality of life gained via a medical treatment. For example, a year lived in perfect health would be a QALY of one, while death corresponds to a QALY of zero. But a year lived with a certain level of chronic pain or limited mobility may be considered a QALY of only 0.5, corresponding to half a year of perfect health. The numerical value of a given symptom or disability is usually determined by surveys; respondents may be asked to consider how many years of perfect health they would be willing to trade for a longer period of survival with a given set of symptoms or disabilities, for example.

The QALY method was developed to quantify the benefits of different health interventions: in particular, health economists calculate cost per QALY to compare the cost efficiency of various treatments. Patients also use informal QALY calculations to decide between options offered by their doctors. For example, a patient with terminal cancer may be faced with the choice between living for one more year with few symptoms versus living for up to five years with cancer treatments that cause unpleasant or debilitating symptoms. QALY can also be used in triage, especially non-emergency triage situations like organ allocation or determining ventilator usage in a pandemic. For example, the policy for allocating lung donations in the US, which until the 2000s was based primarily on how long a patient had been on the waiting list, now considers the patient's likelihood of survival over the course of the year following the transplant.[10]

Rather than simply maximizing the number of lives saved during an emergency, QALY triage attempts to maximize the number of years lived by the survivors, adjusted for quality. This approach requires the medical providers to compare and rank the quality of their patients' lives, which can quickly get ethically murky. Simply comparing *non-adjusted* life years is already a form of age discrimination—does a sixty-year-old deserve to live more than a seventy-year-old? Most people would advocate for saving the life of a child, with so many years of life ahead of them, over the life of an adult. On the other hand, if a space settlement's children are protected at the expense of many adults' lives, what happens when there are not enough adults left to adequately care for the children?

This gets even more complicated when we attempt to adjust this calculation based on quality. How can we possibly objectively compare the "quality" of two people's lives? If we revisit the idea that "each person's desire to stay alive should be regarded as of the same importance and deserving the same respect as that of anyone else," how can we argue that the life of a person living with chronic pain should not weigh as heavily in our consideration as a person living without? Disabled people's desire to stay alive should be given the same respect as the desires of those not living with disabilities, and distributing scarce medical resources using the QALY method treads dangerously close to the ableist notion that we should rank disabled lives as less worthy of saving.

SOCIETAL FACTORS

THE PATIENT AS A MEMBER OF THE COMMUNITY

Both egalitarian forms of triage like a lottery system (in which everyone has an exactly equal shot at receiving medical care) and traditional utilitarian triage (which attempts to maximize the number of lives saved) explicitly ignore the identity of the patients in need of care. But the Ethics Committee of the US Organ Procurement and Transplantation Network (OPTN) argues that justice should be considered equally with utility when considering the problem of organ allocation.[11] A purely utilitarian system of organ allocation, for example, would distribute organs preferentially to patients with a higher likelihood of surviving for at least ten years after transplant. But suppose that a statistical analysis revealed that patients living in poverty were more likely to die within those ten years due to other factors, like poorer nutrition and living conditions, or an inability to pay for medications. A utilitarian approach would insist that wealthier patients should receive preference for organ transplants because they'll be able to live longer with those organs, increasing the number of life years saved by each donated organ. But the OPTN recognizes the injustice of withholding organs from poorer patients purely due to their economic circumstances and notes that such a policy would pile on yet another burden to their reduced life expectancy.

The principle of justice suggests that medical providers in a space settlement may need to be cognizant of certain elements of a patient's identity in order to ensure that care is being fairly distributed. This would allow the providers to consider the role that societal inequalities have already played in the patient's life. But what about the role that the patient plays in society? Should that also be considered a factor in triage?

Philosopher Keith Abney noted that "another way we could think about [triage] is in terms of social value, or mission value. So, if we have to decide between the pilot of the craft versus a passenger, maybe we better save the pilot, right? If we have to decide between the person who's responsible for keeping the atmosphere breathable versus a passenger, well, maybe we better make sure the atmospheric engineer lives."[12] This approach attempts to maximize the number of lives preserved across a whole population, rather than only the population of patients.

As with other types of utilitarian triage, this system of prioritizing patients based on their skillset is also frequently used on the battlefield. As physician Ken Iserson said, "The Germans did this in World War II: they took their front-line soldiers and they tried to patch them up before anybody else. In civilian disasters, it might be the firemen, it might be the physician and nurse, it might be the rescue person who goes into buildings. If they get hurt, try to patch them up so they can maximize the good."[13] Similarly, during the rollout of the COVID-19 vaccines in 2021, many jurisdictions prioritized health care workers for the first available vaccines, to reduce staffing shortages in the strained health care system.

Such an approach requires medical providers to place unequal value on the lives of different members of the community. The OPTN Ethics Committee specifically argues against taking social worth into account when distributing organs, noting that "considering one person more useful to society than another, based on prevailing social values, may be a matter of opinion or good fortune in the random distribution of natural and socially cultivated talents and abilities."[14]

In the US, where the OPTN operates, no single patient is likely to predictably cause or prevent the deaths of a large number of other Americans. Every death of a patient is a tragedy, but no more so than that of any other patient. In the early days of a space settlement, however, there may actually be "indispensable" members, people who essentially hold the community's safety in their hands due to their skills and expertise. The authors of an influential 2009 paper on scarce medical resource allocation noted that prioritizing health care workers during a (hypothetical, at the time) pandemic can be justified by utilitarian arguments: "Prioritising essential health-care staff does not treat them as counting for more in themselves, but rather prioritises them to benefit others. Instrumental value allocation thus arguably recognizes the moral importance of each person, even those not instrumentally valuable."[15] Similarly, prioritizing medical treatment for engineers or pilots in a space settlement could ultimately save more lives than distributing care equally.

On the other hand, such a system could be easy to abuse. For example, community leaders could use their political power to argue that they deserve priority treatment due to their position in the settlement's hierarchy. Even in the absence of this kind of self-serving behavior, a triage system based on instrumental value could also produce indirect injustices. For example, suppose that a settlement's triage policy prioritizes the health of the community's doctors over health care providers with lower levels of training, such as nursing aides who care for the settlement's elderly residents. From a utilitarian perspective, doctors can save more lives in an emergency than nursing aides can. However, in the US, Black Americans are underrepresented as physicians by a factor of 2, and overrepresented as nursing aides by a factor of nearly 3.[16] If this demographic pattern were reproduced in our hypothetical space settlement, Black residents would be less likely to qualify for prioritized medical treatment than white residents. Once again, embracing a purely utilitarian triage model risks reinforcing inequalities that already exist in a society: in this case, the lack of access to higher levels of medical education for marginalized populations.

Of course, if we bring the entire American health care model to space, the settlement may end up using a different metric for distributing medical care, one that is neither utilitarian nor just: the ability to pay. As Keith Abney pointed out, the medical resources that will be scarce in space are not infinite here on Earth: "Medical rationing is, in fact, a fact . . . [It's just that] on Earth, it's usually rationed by ability to pay, so poor people don't get as much as rich people. In space, it's going to have to be more radically rationed, and we're going to need to decide on the basis on which it's rationed."[17]

Abney noted that while NASA doesn't currently ration health care for its astronauts in space based on their ability to pay, the growing private spaceflight industry may approach medical resource distribution differently. "You could easily imagine, if you're going up with Mars One or Elon Musk's SpaceX, that you might be able to pay just a little, and be told, 'You're going to have only minimal medical care on your voyage.' Or you could pay for the elaborate version, the platinum edition, and get much better, guaranteed health care. There are all kinds of possible solutions to this, but we're going to have to think through what we believe to be ethical and what we should allow."

HOW SHOULD WE DISTRIBUTE MEDICAL CARE IN SPACE?

AND WHO GETS TO DECIDE?

For most of the topics covered in this book, I've emphasized the importance of planning ahead, deciding now what kind of systems we'll need in space to ensure that our values are represented. This is doubly important for medical systems, as Ken Iserson said: "It's not, 'Let's make a decision and set up a plan when something happens.' This has to be well planned out, certainly in a space mission, and usually in other long-term missions that have medical components, especially in resource-poor areas or after disasters."[18] One of the most valuable characteristics of a triage system is that the ethical debates and decisions about distributions of care take place *before* the system needs to be used in an emergency. This allows providers to fall back on the objectivity and clarity of the system, avoiding subjective and potentially biased decision-making in the heat of the moment.

But who should develop the triage system to be used in space? This question is a problem even for terrestrial health care systems. Certainly, medical experts like physicians should be included in the process, especially those with experience in treating patients with limited resources. Their education and training provide a scientific basis for understanding the problems of patient assessment and triage, and their experience in the field can help them anticipate what kinds of difficult medical decisions they may be faced with in the future. Medical ethicists can also play a valuable role, given their familiarity with the history of these debates over previous generations and their training in ethical deliberation.

But what about the rest of us? It may well be our descendants who wind up needing medical care in a future space settlement, just as we ourselves may need ventilators or vaccines or some other scarce medical resource on Earth in our own lifetimes. Even if these hypothetical future patients are not related to us, decisions about health care distribution will affect members of the general public, not just physicians and ethicists. Shouldn't the general public have a say in the debate?

Medical policymakers have already been working to increase public collaboration for developing triage protocols on Earth. As bioethicist Robert Veatch points out, health professionals tend to argue for policies that favor efficiency and utilitarianism, while laypeople prioritize justice and equity; to find a balance between these principles, both groups must contribute to the conversation.[19]

For example, in 2013, a team of American physicians and bioethicists performed a pilot study in which members of the lay public were presented with a hypothetical scenario involving a severe flu pandemic.[20] Participants discussed potential options for distributing a limited supply of ventilators to flu patients, considering many of the different approaches discussed in this chapter. The researchers observed that "the participants' engagement in the project was striking"; they not only grasped the technical details of the scenario but were also able to hold complex, nuanced discussions of the ethical implications while fostering an environment of mutual respect for alternate opinions. The study also found that different communities emphasized different principles or concerns during their conversations, indicating that "people engage with medical ethics on the basis of their life experiences, social roles, political concerns, and cultural beliefs."

This kind of deliberative democratic methodology, in which citizens participate in public, informed discussions of policy, can enhance public trust and transparency, both of which are crucial in a health care system. But the results of this pilot study also suggest that the public has much to offer the policymakers, as well, most importantly their lived experience and unique perspectives on these complex ethical questions. As we design the space settlements of the future, we should engage members of the public in discussions of all aspects of policy—not just health care—to ensure that these valuable contributions are not missed.

Of course, including more participants in these debates will certainly not help to simplify matters. "It's hideously complex," Abney admitted. "Ethics is hard. Policy is hard. But we can think through some of these issues and say, 'No, for sure, these are going to be serious problems and you need to try to decide these beforehand, or else things will be disastrous.'"[21]

12

WHICH WAY IS MECCA?

It is late June, the longest night of the year, and you are celebrating Wiñoy Tripantu at your home in southern Chile. You mark the solstice with your family, eating fried pumpkin dough and drinking beverages made from fermented maize, gathering around the bonfire, and listening to the stories of your grandparents. As you welcome the return of the sun and the start of the new year, you feel a connection to your Mapuche ancestors through these ancient traditions.

It is late December, the longest night of the year, and you are celebrating Shab-e Chelleh at your home in Iran. You mark the solstice with your family, eating pomegranates and watermelons, lighting candles, and listening to the stories of your grandparents. As you welcome the return of the sun and the start of the new year, you feel a connection to your Persian ancestors through these ancient traditions.

The year is 2110 CE, and it's been a decade since you left Earth to build a new life in space. The settlement is doing well: your family has grown, your habitat is expanding, and the community is thriving. You want to acknowledge this milestone with a celebration. But you're also a person of faith, and you feel the need to express your gratitude in a quieter, more personal way, too. What can you do? How have you adapted your religious rituals and cultural ceremonies to life in space? How have your beliefs and traditions helped *you* adapt to your new environment?

HOW WILL SPACE CHANGE THE WAY WE LIVE?

Most of the topics covered in this book are focused on the *survival* of the humans living in a space settlement: How will we ensure that everyone has enough air to breathe and food to eat? Will they have access to medical care, safety from violence, freedom from exploitation and oppression? But for intelligent, social animals like humans, life is about more than simply avoiding death and harm. We also seek connection with our fellow humans—our families, friends, and communities—and we form our identities from these bonds. We reinforce these connections through our cultural practices, like cooking a grandparent's favorite recipe, making a pilgrimage to a holy site, or cheering for the hometown sports team. Culture is part of what makes us human, and it connects us not only to each other but also to the generations that came before us and the generations that will follow.

Human cultures are shaped by our physical environments. For example, a culture's traditional attire is heavily influenced by the local climate, which is why the Sámi people of northern Scandinavia developed footwear that included fur-lined boots and the Māori people of New Zealand did not. The cuisine of a given culture depends on what types of native plants, spices, and livestock are available; arts like jewelry and architecture reflect the local region's mineral resources. Space is a radically different environment compared to anywhere humans have lived on Earth, and this is sure to have a profound impact on the culture of future generations living in space. For example, the cuisine, clothing, and art in a space settlement will depend on the types of materials available to the residents, and dance, sports, and architecture will be constrained or liberated by the settlement's distinctive gravity.

Beyond these physical influences on human culture, the space environment may also create less-predictable, psychological effects. For example, the perspective provided by living in space has been known to cause cognitive shifts in individual space travelers, a phenomenon known as the "overview effect." Coined by Frank White, the term refers to the sudden awareness of humanity's unity and Earth's fragility experienced by some astronauts when they first view the Earth from space.[1] It's unknown yet whether individual human travelers to Mars will experience similar or even greater shifts in perspective, but at the population level, this effect would also likely drive cultural change.

While the space environment is sure to shape the culture of the humans who build a new society there, the first generation of space settlers will also bring their own cultures with them from Earth. How will these various terrestrial cultures mingle and evolve as we build new communities together in space? How will our cultural attitudes and practices help us survive the harsh realities of an alien environment?

HOW CAN CULTURE IMPROVE OUR LIVES IN SPACE?

LESSONS FROM ARCTIC LIVING

Humans have been visiting space since 1961 and living there continuously since 2000. What kind of culture have we brought to space so far? Naturally, the culture of space travelers has been representative of the nations capable of human spaceflight, dominated by the two major space superpowers, the US and Russia. More specifically, the people who work in space have historically come from just a handful of fields and industries, each with their own distinct cultural attitudes and behaviors: early astronauts were military test pilots; many payload specialists today have scientific backgrounds; and now tech entrepreneurs are joining the mix. While the diversity of cultures represented in space is growing as access to space increases, there are many cultures on Earth that are currently poorly represented in the astronaut corps but could nevertheless provide valuable lessons on how to build long-term, sustainable, and thriving communities in space.

Alice Qannik Glenn has been making this point on behalf of the Iñupiat people of northern Alaska for several years. Glenn is an Iñupiaq woman herself, born and raised in Utqiaġvik, Alaska, the northernmost town in the United States. While studying aerospace science at university, Glenn noticed a number of similarities between her people's Arctic home and the austere environments that space settlers will one day attempt to inhabit.

"Everything that we were learning in our book about all of these challenges in space, it kind of reminded me of back home," Glenn said.[2] She later spoke at the Icarus Interstellar Starship Congress in 2017 about how her people have adapted to living in the Arctic, noting that long-term human spaceflight and space settlement planners could look to societies like the Iñupiat for lessons on how to build thriving communities in an isolated and deadly environment.

I asked Glenn to describe some of the similarities between the challenges of living in the Arctic and living in space. Glenn calls these "perceived challenges," because "although they're recognized as problems for astronauts in space, some of these are just facts of life for Iñupiat people." The first and most obvious of these perceived challenges is the cold, dark, and desolate landscape. In the northern latitudes of the Arctic where the Iñupiat live, the sun can set for weeks at a time, and the temperature can plummet to -60°F. A visitor from the mainland United States can feel as if they've been dropped on the moon, Glenn said.

Another similarity with space is the relative isolation of Arctic communities. While rockets aren't required for visiting the Arctic, many villages in Northern Alaska are only accessible by airplane for most of the year, and even then, flights are subject to the weather. Communications with the rest of the world can be intermittent, due to poor cell network coverage and internet infrastructure. Even the comfort and shelter of Arctic homes can create their own problems of claustrophobia and cabin fever. "When it's cold and dark outside, a lot of people tend to stay inside and stay warm, and don't want to brave the cold," Glenn said. "So I think that's similar to living on a spaceship for an extended period of time."

So, how have the Iñupiat learned to live with these challenges so well that they see them as simply a part of life? Glenn said that the first and most important characteristic of her culture is adaptability. "Iñupiat culture, economy, and the environment are ever-changing, and the people have learned to adjust to external forces, just from thousands of years of living there, to colonization, and now climate change. So I think that's the biggest asset we have, is adaptability."

Similarly, Glenn noted, Iñupiat people understand the importance of preparedness. "We know that it's going to be cold and harsh outside, so we need to be prepared for it. That's just all you can do, really. And you need to be prepared for white-out conditions, snowstorms, all of those things." Adaptability and preparedness are already crucial traits for space travelers today, and will be even more so for space settlers, who will need to be able to solve unanticipated problems in brand-new environments with limited resources and help.

Space settlers will need more than just individual problem-solving skills to build healthy societies. Today's space exploration is a team effort, and the ability to get along with your neighbors will be even more important in the space settlements of the future. "Attitude is everything [in the Arctic]," Glenn said. "You can't just complain about the darkness or the cold. Nobody does that, because who are you going to complain to? The person that lives there, too? You're not going to get anything done. You're not going to convince anybody of anything. It's just wasted time, you know?"

The sense that "we're all in this together" has helped countless groups of humans survive miserable conditions throughout history, including wars, famines, and pandemics. Glenn explained that the Iñupiat also recognize the importance of cooperation and community for surviving in the Arctic. "Everybody knows that the community is better than an individual. [Having] more people is better than just one and the more help that you have, the better you're off. Nothing big gets done alone."

Researchers in space medicine and long-term human spaceflight are already concerned about the potentially damaging psychological effects of the darkness, stress, and isolation of space. Iñupiat culture protects against these mental health stressors with social activities and celebrations. In the Arctic, Glenn said, "You don't stay home. Nobody is alone all the time. It's not healthy. Iñupiat people know that, so they have a bunch of gatherings." Glenn described several of these celebrations to me, including the Christmas Games; the Kivgiq, a messenger feast celebration held every three years; and Nalukataq, the shared celebration of the end of the springtime whale harvest. Communities in space will likely develop celebrations and holidays of their own, in addition to the traditions they bring with them from Earth.

Glenn explained that her people also draw strength from their connection to their community, their ancestors, and their land. "Although you may feel or look like you are alone, in this vast Arctic landscape, there's just the knowledge, knowing that your people have lived there for thousands of years and it's not the first time, that you're not the only person to have stepped on this ground." Perhaps this is what our descendants will feel when they look out across a bleak Martian landscape or into the vast abyss of space: a sense of connection, rather than isolation.

The Iñupiat are not the only people we can learn from as we begin to build communities in space. But we can't simply cherry-pick the most useful characteristics of other cultures for our own use, colonizing their traditions and beliefs the way European colonizers took other people's land, resources, and labor for themselves. Instead, we must ensure that members of all cultures on Earth have equitable access to space, both in terms of space travel itself and of the decision-making processes regarding humanity's future in space.

HOW WILL WE ADAPT OUR CULTURAL PRACTICES TO SPACE?

CELEBRATING LIFE AND DEATH AWAY FROM EARTH

A key component of making space accessible to everyone is to ensure that every resident of a settlement feels free to practice their culture. This is already difficult for humans on Earth who live in communities where their culture is the minority. Space will not solve this problem; cultural conflicts will continue to emerge as people from different backgrounds on Earth attempt to build settlements together or near each other in space. Preventing these conflicts from evolving into oppression or erasure will take conscious, deliberate effort.

Even the space environment itself could create obstacles for people trying to reproduce their cultural practices away from Earth. Consider all the basic physical phenomena we take for granted on Earth that will be scarce or absent in space, like the gravity field we evolved in, regular sunrises and sunsets, and easily obtained water, air, and even dirt. Many of these characteristics are integral to cultural practices of human societies on Earth.

For example, Christian theologian Michael Waltemathe of Ruhr University described some of the problems with adapting death rituals for space. The diversity of burial rites among human cultures shows that these practices are more than simply techniques for disposing of corpses hygienically; they reflect our beliefs about humankind's place in the physical world and our relationship with each other. Many (though not all) human burial practices involve returning the remains to nature. For example, bodies interred in the ground without embalming or mummification will eventually decompose into the surrounding soil, while Tibetan sky burials place the body on a mountaintop to be consumed by scavenging animals. In space, the resources contained within a deceased human body, like water, calories, and nutrients, will be too scarce to waste by removing them from the ecosystem, but scavenging wildlife and microbe-rich graveyards will also be hard to come by.

"Now on Earth, you can just bury someone in the ground, and we all know that it is repurposed by organisms back into the ecology," Waltemathe said. "And you would probably need to do that in a self-sustaining habitat as well, but there is sort of a difference between putting somebody in a recycler and putting someone in the ground, right?"[3] During trips between planets, will spaceship crews practice pseudo-naval space burials, ejecting the deceased out into the void of space like in *Star Trek*? Will they take advantage of the surrounding vacuum to flash-freeze and then crumble the body for easy transport, a concept that author Mary Robinette Kowal explored in her *Lady Astronaut* series? Will we soften the common human taboo against cannibalism if a community's food chain becomes significantly shorter between the bodies of the recently deceased and the nutrients feeding the still-living? How will we adapt our cultural attitudes and beliefs about death to these practical challenges?

Religious rituals might be especially tricky to reproduce in space since there are often strict rules about how and when rituals should be performed. How will settlers adapt their religious practices for the space environment without violating the tenets of their faith? This is not a new question, given that people of faith have been traveling to space for decades now. Modern Americans, raised on decades of public debate pitting science versus religion, often assume that space workers, from astronauts to astronomers to private spaceflight entrepreneurs, are predominantly atheistic, or at least nonreligious. But anthropologist Deana Weibel, who studies the intersection of space and religion at Grand Valley State University, has found that many space workers not only hold some kind of religious belief but also feel that their work is motivated by their belief system.

"What I find interesting is that when I sit down and start talking to people, the evangelical Christians are *absolutely* represented in space exploration," Weibel said. "And [for] those who are really into it, [their] religion really shapes the way that they understand it."[4] While some of the Christian space workers she interviewed expressed a need to hide their beliefs from their colleagues, Weibel found that religious space workers from other traditions saw no conflict between their faith and their scientific work. For example, Weibel interviewed a young Hindu space worker from a Vedic tradition: "It was all about energy," Weibel recalled, "and energy from the universe going out in certain ways, and how everything she knew from physics was completely compatible with the Hindu perspective of how the universe started, patterns of energy. Her family were Brahmin, and her male relatives were all scientists and professors who also had extremely important ritual roles in the temple."

Weibel observed that astronauts in particular tend to be more openly religious, possibly due to the large number of astronauts with military backgrounds. Religious activities by American astronauts have generated some controversy in the past, because these practices clash with NASA's status as a government agency in a nation that values the separation of church and state.

"When Apollo 8 flew to the Moon and back, and they orbited the Moon, they read from biblical scripture on Christmas Eve, 1968," Waltemathe said. "They basically read the story of creation, the first four verses, and then they blessed the population of Earth. And that caused quite a stir, because a couple of days later, the American Atheist Association sued the U.S. government for spending taxpayers' money on religious messages from outer space."[5] Ultimately, the Supreme Court dismissed the case because lunar orbit is not in their jurisdiction.

Some religious rituals translate well into the alien environment of space, or at least as well as other everyday human activities. Waltemathe recounted that Apollo 11's Buzz Aldrin brought wafers, wine, and a silver chalice to celebrate Holy Communion on the Moon with Neil Armstrong, a ritual that "worked very well in lunar gravity."[6] But other practices are more closely tied with our home planet. For example, practicing Muslims are required to pray five times a day, and these prayers have a specific format. They should be done in a clean place and involve standing, bowing, or prostrating on the ground, all while facing toward the holy city of Mecca, in Saudi Arabia. How can Muslim space travelers perform these prayers in microgravity? What if they're on board a space station like the ISS, orbiting the Earth once every ninety minutes?

These types of questions are not new in Islam, Muslim theologian Mehmet Ozalp of Charles Sturt University explained; in fact, they predate space travel. As Muslims have travelled far from their homeland over the years and adapted to new technologies, they've had to adapt their practices as best they could. "What do you do if you're on a ship and the direction of the ship is changing all the time? [Islamic scholars] have come up with solutions to this," Ozalp explained. "They say that you start off facing in the direction of Mecca when that ship is in that direction at that particular time." These questions arose again with the invention of airplanes, "Especially when you're in the Pacific, where you could [face] both directions. And then there's the issue of curvature of the Earth. So all of this comes into play in space, as well."[7] Fortunately, Muslim astronauts can take advantage of the wisdom of those earlier Islamic scholars: an astronaut in orbit can start their prayers facing Earth, in the direction of Mecca if possible, even if their spacecraft continues to move or turn during the prayer.

"Islam is a flexible religion, inherently, in its practices," Ozalp said. "Even from the time of the Prophet Muhammad, for instance, when Muslims pray, or before praying, they have to have ablution. And what if you don't have water? What are you going to do? There's something called sand-washing: you tap your hands on sand and you kind of wash your limbs. So there's flexibility."

A number of religions, including Islam, use timekeeping methods that rely on the Earth's position relative to the sun, a useful technique for rituals that were developed before clocks were invented. For example, the Jewish Sabbath begins at sunset; many Buddhist holidays fall on the day of the full moon; and the dates of the spring and winter equinoxes are celebrated in many cultures and religions. The five daily prayers in Islam are to be performed at dawn, midday, afternoon, sunset, and night. Even before space travel, concerns about using celestial timekeeping methods arose on Earth, where the timing of sunrise and sunset depends on one's distance to the equator.

"This question came up for Muslims in Sweden or Norway, where there is six months with no sun," Ozalp explained. "And the solution is that you either assume that you are in Mecca and you follow Meccan times (a 24-hour cycle), or you go with the nearest normal day, a latitude where it would give you a reasonable day where you can have five proper, daily prayers. So, if you were on a space station, that's what you would do."

Malaysian astronaut Sheikh Muszaphar Shukor was confronted with these questions when he traveled to the ISS in 2007, so Malaysia hosted a conference of Islamic scholars to determine requirements for prayer and fasting for Muslims on the ISS. The document they produced describes when to pray and fast (using the time zone of launch), which way to face (toward Earth, or Mecca if possible), and how to perform the physical postures in a weightless environment. The Malaysian National Fatwa Council also concluded that dry ablution could be performed on the ISS by "striking both palms of hands on a clean surface such as a wall or mirror," even without any sand or dust.[8]

"Fundamentally," Ozalp said, "the purpose is to pray. As the Quran says, God is not in the West, not in the East. God is beyond space and time."

HOW WILL SPACE CHANGE US?

GAINING A NEW PERSPECTIVE

Many space settlement evangelists hope that living in space will change our society for the better (or at least, the portion of our society that moves off Earth). Many of these claims are based on the physical separation that will exist between future space settlements and the "old world" of Earth, which will provide the opportunity to experiment with different social, legal, and economic systems away from the influence of terrestrial society. Over time, this separation will naturally create differences between humans living in space and on Earth, as our traditions, politics, and even languages evolve in distinct environments. But besides the isolation from the rest of humanity, are there unique aspects of the space environment that might fundamentally change the way we live?

The so-called overview effect, for example, has been known to change an individual's conception of humanity's place in the universe. The ability to see the entire Earth at once emphasizes our planet's beauty and fragility and reveals the artificiality of human borders. Anthropologist Deana Weibel's interviews with astronauts have also revealed an experience she calls the "ultraview effect," a sense of smallness or incomprehension that can occur when viewing space from a position outside the Earth's atmosphere. However, Weibel has also observed that these effects are not universal among space travelers: "I didn't really feel anything," one pseudonymous astronaut told Weibel during an interview. "It's kind of a letdown . . . It was a beautiful sight and a unique vantage point, but there was nothing about it that I felt in any way unlocked any kind of philosophical mysteries or spiritual mysteries."[9]

It's unclear how much these psychological effects of space will shape future human cultures. Will the new perspectives provided by long-term, long-distance space travel shift human societies away from nationalism, environmental exploitation, and intraspecies conflict? Or will we simply draw new borders in space: between mining zones on Mars, between space station habitats in orbit, or even between space settlers and their kin back on Earth?

Space travel is so inaccessible to most humans on Earth today that our culture places a heightened value on people or even objects that have travelled to space and returned. But as we extend our civilization beyond this planet, the opposite will occur as we transform objects in space through our presence and use. Unremarkable piles of dust sitting on the surface of the Moon today may one day form the wall of someone's childhood home. Chunks of ice frozen on Mars's north pole might water the first crops in humanity's future space settlements. While the space environment will change human settlers individually and as a society, humans will also bring meaning and sentiment to the physical worlds where we build our homes.

As an illustration of this phenomenon, Weibel described her interview with Jewish rabbi Shaul Osadchey, who helped astronaut Jeff Hoffman bring a tiny Torah to space with him on the space shuttle *Columbia*. "So do you think bringing the Torah into space made it more holy?" Weibel asked the rabbi. "No," Osadchev replied. "If anything, it made space more holy."[10]

PROVIDING JOY TO OUR DESCENDANTS IN SPACE

Humans living in space will create new cultures influenced by their extra-terrestrial environment as well as the unique combination of backgrounds of the settlers who choose to migrate from Earth. We'll invent new holidays in space, tied to new sunrises and sunsets. Our art will depict alien landscapes that are no longer alien to the artists, and we'll create sports and dance styles impossible to recreate in Earth's gravity. But our descendants in space will also preserve some of the traditions and values of their terrestrial ancestors, a bridge to a home planet they may never visit.

The culture of the first generation of space settlers, then, will have a profound effect not only on the immediate success of a settlement but on the lives of their descendants far in the future. If we want those descendants to live fulfilling, joyful lives, we can help set them on the right path by ensuring that the cultural values and attitudes we bring to space are those with the proven ability to help humans thrive in harsh and dangerous environments here on Earth. How can we identify these characteristics? When I put this question to Alice Qannik Glenn, she emphasized the importance of including diverse cultures in space. The Iñupiat certainly have much to teach the rest of us about finding joy and resilience in a cold, dark, isolated environment—What else might we learn by welcoming more voices into the conversation? "If I were able to go to space as an Alaskan Native Iñupiaq woman, I would know some of these things," Glenn said. "I would know some answers that someone else might not know, or they might know something that I don't know. Having a diverse group of people included in the conversation, and in space—I think that's just paramount. Because you can't be sending all the same people to space and expecting to come up with new ideas."[11]

To preserve our heritage as we move into space, we'll also need to adapt our cultural practices to our new environment, just as we'll need to adapt our agriculture, technology, and medical care. Mehmet Ozalp noted that Muslims, for example, will need to continue the discussion they've begun about what practicing Islam in space will look like.

"We need to first develop a theology of space and life in space," Ozalp said. "You would look at all the relevant sayings of Prophet Muhammad related to life in foreign places or strange lands or even space . . . So to read these sources with that in mind and bring all those together and develop a theology of what it means to be humans traveling to outer space and living there—that is required, in my opinion. And then that would be the guiding theology that would be of import to addressing individual issues and difficulties that people might face."[12] This is, in effect, what space ethicists have been doing in their own field: combing through the academic literature of philosophy, applying the wisdom they find there to the context of space, and developing a philosophy of what it will mean to be humans living in outer space.

The success of our cultural preparation for space settlement will be harder to measure than our efforts concerning population control or the prevention of war. After all, there are no "right" or "wrong" cultures on Earth, and there are countless ways to build a society in space that will provide a safe and happy environment for our descendants. But if we hope to use space to preserve our diverse and unique human cultures as well as our DNA, we should make sure we're providing future generations with both the ability to celebrate the traditions of their terrestrial ancestors and the freedom to create a new kind of human society with each other.

13

WHAT CAN WE DO
TO PREPARE?

The year is 2023 CE, and you are reading a book about the ethical challenges
of space settlement. You don't agree with everything you've read, but some of
the chapters have raised questions you'd never considered about what life will
be like in space. You want to learn more: to study similar cases from history
and our modern terrestrial societies, to discuss these issues with other people
interested in space settlement. Most importantly, you want to take action, to
actually help shape humanity's future in space. What should you do next?

MOTIVATION

NOW WHAT?

The previous dozen chapters have given you a whirlwind tour of the potential pitfalls we may encounter as we build permanent human communities in space. I've pointed out areas where conflict could arise between
groups of space residents with different goals, where there could exist
opportunities for exploitation and human rights violations, and where
parallels to historical cases encourage us to learn from them as cautionary
tales. This exercise has undoubtedly revealed much about my own political opinions and priorities, not to mention the influence of my personal
background and the culture in which I was raised. In the same way, your
position on these issues is likely deeply connected with your own values
and beliefs.

Perhaps you agree with everything I've written, both the potential areas of conflict in space and the solutions I've proposed. More likely, you don't; each of us has our own concept of what a "good" society looks like. Either way, I'm sure that as you've considered the complexities of building new communities in space, you've thought of questions and scenarios that I haven't considered in this book. Maybe you're especially concerned about one or two of the topics I've discussed and you'd like to make sure that space settlement planners consider these particular issues as they move toward their goals. The next obvious question for all of us is: What should we do about it? What actions can each of us take to help build a better future in space?

WHAT CAN STUDENTS AND RESEARCHERS DO NOW?

THERE'S ALWAYS MORE TO LEARN

In a typical scientific research article, one of the last sections is usually dedicated to discussing the future work left to be done on the project. The authors will propose gathering more data or investigating possible lines of research suggested by the article's results. Certainly, there is plenty of work left to do on the topics discussed in this book, but there are also entire fields of research that can already contribute significantly to conversations about human rights in space, and readers interested in becoming more active in the issues I've raised should begin by exploring these fields further.

For example, the history of human migration across the Earth (and, later, the colonization of already-inhabited territory) can provide us with examples of the difficulties and dangers of building new communities in distant lands. Most of the historical examples of colonialism in this book have focused on the European colonization of the Americas, particularly North America, given its outsized effect on the mythology and culture of today's American-dominated space industry, but there is more to learn outside these continents, particularly in regions and eras understudied by Western historians. Along with history, fields like sociology, anthropology, and political science analyze the ways in which humans tend to form societies and interact with each other. The insights provided by these fields can help us predict how we'll continue to create societies in a new environment like space and where we're likely to run into trouble.

There are also numerous subfields of ethics that will be relevant in space. Ethics is a branch of philosophy that provides useful tools for the kinds of large-scale decision-making that will be required for space settlement. The broad field of bioethics, for example, includes the questions of medical and reproductive ethics that doctors, patients, parents, and children will struggle with in resource-poor space settlements. Environmental ethics examines our relationship with the world around us and can help us clarify our priorities regarding the distribution of space resources and our responsibilities to the physical environment of space and any potential life there.

We also have much to learn from the activists who have fought for justice on Earth, in the past and today. For example, champions of environmental justice work for the preservation of species, the formation of legal protections like national parks, and the defense of Indigenous land and resources. Labor rights and economic justice activists seek to protect less powerful members of our economic hierarchy from exploitation and abuse. Transformative and restorative justice advocates challenge our current criminal justice system to push us toward better methods for addressing harm in our society. Certainly, these activists can educate all of us about the structure and values required for a just society but they also provide a roadmap for how to raise awareness of and act on social justice issues in space settlement.

While all these terrestrial-focused fields can be applied to space, the field of space law is a well-developed area of study that directly examines our legal structures in the context of space. International treaties, national laws, and regulations governing space have existed for decades; the United Nations Committee on the Peaceful Uses of Outer Space was formed the same year that NASA was established, 1958, to ensure that space is used by nations for peaceful purposes that benefit all of humankind. The recent growth in privately funded space activities, as well as increasing tensions between space-capable nations like the US, Russia, and China, continues to draw attention to the questions left to be answered about how we'll govern ourselves in space.

Space ethics itself is also a well-established field. According to two of today's leading space ethicists, James Schwartz and Tony Milligan, the field emerged from discussions in the 1980s about environmental ethics, planetary protection, and the hypothetical terraforming of Mars, although philosopher Hannah Arendt was examining the implications of space rhetoric about "escaping men's imprisonment to the earth" as early as 1958.[1] Today, the field of space ethics encompasses everything from modern space exploration to potential future settlement. Space ethics is often combined with perspectives from fields like sociology and anthropology to form a broader, interdisciplinary community that has been labelled with a variety of acronyms, such as ELSI for the "ethical, legal, and social implications" of space, or STS for "science, technology, and society."

Space anthropologists like David Valentine, Valerie Olson, and Debbora Battaglia consider space a "crucial site for examining practices of future imaginings in social terms, and for anthropological engagement with these practices."[2] For example, Kathryn Denning uses an anthropological perspective to suggest how we could prepare for the consequences of the detection of extraterrestrial life, while Michael Oman-Reagan's "Queering Outer Space" encourages us to "water, fertilize, and tend the seeds of alternative visions of possible futures in space."[3] Space historians, meanwhile, analyze the history of humanity's interaction with space from various perspectives, as in Lisa Ruth Rand's environmental history of Earth orbit or Asif Siddiqi's study of the global history of space exploration beyond the dominant US and Russian narratives.[4] Political scientist Daniel Deudney and architect Fred Scharmen both published books in 2021 probing the motivations behind today's space settlement enthusiasts from the perspectives of their respective fields of expertise.[5] All these examples barely scratch the surface of the numerous experts in diverse fields who are currently working toward a better understanding of the problems we face in space.

There's growing interest in interdisciplinary research in the US academic system, motivated by the benefits of bringing multiple perspectives to a problem. But during our discussion about Indigenous perspectives on environmental justice, University of Michigan professor Kyle Whyte pointed out that while the Western worldview draws a distinct line between scientific and social fields, Indigenous knowledge systems never created this distinction. "For a lot of scientists [who] are just now trying to get into interdisciplinarity and transdisciplinarity, I think it's hard for them to understand that a lot of other knowledge systems did not have that," Whyte said. The Western system that dominates conversations about space, he says, has "this weird hang-up about objective and non-objective, or objective and social, or objective versus cultural. But not all knowledge systems had that hang-up, so there's no need in an Indigenous science to 'move toward interdisciplinarity,' or transdisciplinarity, because there wasn't a division of the disciplines in the first place."[6]

In the process of writing this book, I've been pleased to discover how eager researchers in non-STEM humanities fields have been to share their work, even those who never think about space and were initially startled to be approached by an astrophysicist. If you're a STEM student or researcher just beginning to think about these issues, consider that many of these topics might already be under discussion in humanities fields that you may have never even taken an introductory class in. I encourage you to reach out to researchers in these fields, just as you'd collaborate with fellow scientists in adjacent subfields of your own area of study, to help rebuild the interdisciplinarity that we've lost in Western science.

WHAT CAN ENTREPRENEURS AND POLICYMAKERS DO NOW?

TAKE TIME FOR SELF-REFLECTION

Today's private space companies and governmental space policymakers can have a significant effect on the space settlements of the future by putting into place not only legal structures to regulate space activities but also cultural norms. If today's space decision makers make a point of publicly prioritizing human rights and ethics in their work, these values will have a better chance of persisting through the coming generations of space workers and, eventually, space residents.

When I first started researching space ethics and human rights, I was disappointed by the number of big players in the space industry who seemed dismissive of, or, at best, ignorant of these issues. Since then, however, I've had the chance to talk with many people in the private space industry and public space agencies who *are* eager to buck this trend and further incorporate justice and ethics in their work, but don't know how.

If you find yourself in this group, my first suggestion is to go back and re-read the previous section, then reach out to the experts in relevant fields for advice, even those who work primarily in terrestrial contexts. There are also organizations whose goal is explicitly to connect interested parties in the space industry with researchers and experts in fields like ethics, labor rights, and environmental justice, including my own nonprofit, the JustSpace Alliance, which I co-founded with astronomer Lucianne Walkowicz in 2018.

Besides seeking advice from outside experts, I recommend examining the strengths and weaknesses of your own organization. How diverse is your workforce, both in terms of expertise (have you considered adding an ethicist to your staff?) and in terms of cultural background and lived experience? As I've noted in multiple chapters, restricting our conversations about space to only one group of people—STEM-educated, predominantly white, predominantly male Westerners, for example—leaves gaps in the kinds of problems and solutions we can imagine for space. If you struggle to find applicants with diverse backgrounds and identities, consider what structures in your community and attitudes in your own organization are creating barriers to those applicants.

Finally, ask yourself what kinds of injustices your organizations are condoning or perpetuating here on Earth today, right now. How do you treat your employees? What methods do you have in place to address interpersonal harm within the organization? Do you know the history of the land your organization occupies? What are the environmental implications of your work, even the indirect ones? Does your organization exemplify the kinds of values that you hope we will one day carry with us to space? If not, there's no need to wait until then; do the work now to push your organization toward a state you can be proud of here on Earth, and you'll create a strong ethical foundation for your future activities in space.

WHAT CAN ALL OF US DO NOW?

A BETTER WORLD IN SPACE AND ON EARTH

Most of my readers are probably not in positions of authority in the space industry, but that doesn't mean you don't have the ability to shape humanity's future in space. Educating yourself is certainly an important first step, and one that should be an ongoing process. The next step is to share what you've learned, and your own ideas that you've developed through your research, with other space enthusiasts. Space settlement advocates, like any group with a shared passion, form communities both formal and informal to discuss their interests and plans. Talk to your fellow space enthusiasts, and the leaders of these communities, and tell them what kind of action you'd like to see to address potential ethical pitfalls as we continue reaching beyond our home planet.

Most importantly, take the time to imagine the kind of world you want to live in. If you truly were inventing a new society from scratch, what would it look like? Who would you welcome into this community? How would the inhabitants spend their days? When things go wrong, how would your people join together to address the problem?

This exercise is a valuable one even if humanity never leaves this planet. If you can picture a better world in space, why not on Earth? If you're willing to fight for the health and happiness of the hypothetical residents of a future space settlement, what's stopping you from fighting for the rights of your neighbors today? Why wait for the rockets to be built? Let's take what we've learned from this exploration of how to help humanity thrive in space and use it to create a civilization that deserves to spread itself to the stars.

ACKNOWLEDGMENTS

I am deeply grateful to all my interview subjects, including my podcast guests and later book interviewees, for sharing their time and knowledge freely with me, especially Karen Backe, the subject of my very first interview. And to all the listeners who accompanied me on my journey: thank you, and I hope you enjoy this continuation.

Thank you to the people who helped this book along its path to publication: my graduate advisor Marc Kuchner, who recommended my work to publishers in his network even after I was no longer his student; Jeff Dean, the first editor to believe in the project, who helped me polish my book proposal and sample chapter; and Katie Helke, my editor at the MIT Press, for guiding me through the publication process and saying very kind things about my writing ability in the meantime.

And of course, thank you to everyone else in my life who encouraged me along the way: to Lucianne Walkowicz, for making me braver; to Dan Dixon and my colleagues at Giant Army, for their patience and flexibility regarding my work schedule; to my crewmates at the Odenton Volunteer Fire Company, for their encouragement and support as I wrote much of this book on shift with them. Thank you to my parents and siblings, for their love and enthusiasm; to Korey Haynes, for her eternal and unrelenting support of my writing; and to Brendan Hayward, for his constant presence and support, his book recommendations, and his very wise advice regarding restraint.

And finally, thank you to every activist who has worked, and is still working, to leave this world better than they found it. This book is dedicated, with gratitude, to every human planting trees in whose shade they will never sit.

NOTES

PREFACE

1. Episodes and transcripts of the *Making New Worlds* podcast are available at http://www.makingnewworlds.com.

INTRODUCTION

1. Wendell Berry, "Comments on O'Neill's Space Colonies," in *Space Colonies*, ed. Stewart Brand (San Francisco: Waller Press, 1977), https://space.nss.org/settle ment/nasa/CoEvolutionBook/DEBATE.HTML.

CHAPTER 1

1. Michael D. Griffin, "Chapter 1," in *NASA at 50: Interviews with NASA's Senior Leadership*, ed. Rebecca Wright, Sandra Johnson, and Steven J. Dick (Washington, DC: US Government Printing Office, 2012), 21, https://www.nasa.gov/pdf /716218main_nasa_at_50-ebook.pdf.

2. Robert Zubrin, *The Case for Mars: The Plan to Settle the Red Planet and Why We Must* (New York: Touchstone, 1996).

3. Konstantin Tsiolkovsky, letter to Boris Vorobiev (1911). See, e.g., Jacques Arnould, *Icarus' Second Chance: The Basis and Perspectives of Space Ethics* (New York: SpringerWienNewYork, 2011). A more direct translation from the Russian would be, "The planet is the cradle of the mind, but one cannot live in the cradle forever."

4. Albert A. Harrison, "Russian and American Cosmism: Religion, National Psyche, and Spaceflight," *Astropolitics* 11, no. 1–2 (2013): 25–44, http://doi.org/10.1080 /14777622.2013.801719.

5. Linda Billings, "Should Humans Colonize Other Planets? No," *Theology and Science* 15, no. 3 (2017): 321–332, https://doi.org/10.1080/14746700.2017.1335065.

6. Deana Weibel, interview with the author, April 2, 2020.

7. For example, Roger D. Launius, "Escaping Earth: Human Spaceflight as Religion," *Astropolitics* 11, no. 1–2 (2013): 45–64, http://doi.org/10.1080/14777622.2013 .801720.

8. Weibel, interview.

9. Charles Wohlforth and Amanda R. Hendrix, *Beyond Earth: Our Path to a New Home in the Planets* (New York: Knopf Doubleday, 2016), 226.

10. Carl Sagan, *Cosmos: A Personal Voyage*, episode 7, "The Backbone of Night," aired November 9, 1980.

11. For example, Panos Roussos, Stella G. Giakoumaki, and Panos Bitsios, "Cognitive and Emotional Processing in High Novelty Seeking Associated with the L-DRD4 Genotype," *Neuropsychologia* 47, no. 7 (2009): 1654–1659, https://doi.org/10.1016/j.neuropsychologia.2009.02.005.

12. Chuansheng Chen et al., "Population Migration and the Variation of Dopamine D4 Receptor (DRD4) Allele Frequencies Around the Globe," *Evolution and Human Behavior* 20, no. 5 (1999): 309–324, https://doi.org/10.1016/S1090-5138(99)00015-X.

13. Gerard K. O'Neill, *The High Frontier: Human Colonies in Space* (New York: William Morrow, 1977).

14. Carl Sagan, *Pale Blue Dot: A Vision of the Human Future in Space* (New York: Random House, 1994).

15. "NSS Statement of Philosophy," National Space Society, accessed March 29, 2020, https://space.nss.org/nss-statement-of-philosophy/.

16. Jeff Foust, "The Cosmic Vision of Jeff Bezos," *SpaceNews*, February 25, 2019, https://spacenews.com/the-cosmic-vision-of-jeff-bezos/.

17. Zubrin, *The Case for Mars*.

18. James S. J. Schwartz, "Myth-Free Space Advocacy Part II: The Myth of the Space Frontier," *Astropolitics* 15, no. 2 (2017): 167–184, http://doi.org/10.1080/14777622.2017.1339255.

19. Martin Robbins, "How Can Our Future Mars Colonies Be Free of Sexism and Racism?," *The Guardian*, May 6, 2015, https://www.theguardian.com/science/the-lay-scientist/2015/may/06/how-can-our-future-mars-colonies-be-free-of-sexism-and-racism.

20. Linda Billings, "Space Cowboys: How Jingoism Corrupts American Rhetoric on Human Spaceflight," *Scientific American* 313, no. 2 (2015): 12, http://doi.org/10.1038/scientificamerican0815-12.

21. J. B. Bury, *The Idea of Progress* (London: Macmillan, 1920), 2–4, https://archive.org/stream/ideaofprogressinooburyuoft#page/xvi/mode/2up.

22. K.-P. Schröder and Robert Connon Smith, "Distant Future of the Sun and Earth Revisited," *Monthly Notices of the Royal Astronomical Society* 386, no. 1 (2008): 155–163, https://doi.org/10.1111/j.1365-2966.2008.13022.x.

23. Charles Bolden, "Sending Humans to Mars," forum, Humans 2 Mars Summit, Washington, DC, April 22, 2014, video, 29:12, https://www.c-span.org/video/?318982-1/human-mars-summit.

24. Sylvia Hui, "Hawking: Humans Must Spread Out in Space," *Associated Press*, June 13, 2006, https://www.washingtonpost.com/wp-dyn/content/article/2006/06/13/AR2006061301185_pf.html.

25. Elon Musk, "Making Humans a Multi-Planetary Species," *New Space* 5, no. 2 (2017): 46–61, https://doi.org/10.1089/space.2017.29009.emu.

26. Ross Anderson, "Exodus: Elon Musk Argues that We Must Put a Million People on Mars If We Are to Ensure that Humanity Has a Future," *Aeon*, September 30, 2014, https://aeon.co/essays/elon-musk-puts-his-case-for-a-multi-planet-civilisation.

27. William K. Hartmann, "Space Exploration and Environmental Issues," in *Beyond Spaceship Earth: Environmental Ethics and the Solar System*, ed. Eugene C. Hargrove (San Francisco: Sierra Club Books, 1986), 122.

28. Kelly C. Smith, "*Homo reductio*: Eco-Nihilism and Human Colonization of Other Worlds," *Futures* 110 (2019): 31–34, https://doi.org/10.1016/j.futures.2019.02.005.

29. Adam Potthast, "Alien Attacks, Hell Gerbils, and Assisted Dying: Arguments Against Saving Mere Humanity," *Futures* 110 (2019): 41–43, https://doi.org/10.1016/j.futures.2019.02.008.

30. Linda Billings, interview with the author, February 27, 2020.

31. Billings, interview.

CHAPTER 2

1. John F. Kennedy, "Address at Rice University on the Nation's Space Effort," speech, Rice University, Houston, September 12, 1962, transcript and video, 18:27, https://www.jfklibrary.org/learn/about-jfk/historic-speeches/address-at-rice-university-on-the-nations-space-effort.

2. Zuleyka Zevallos, "Ep 01: Why Are We Going?," interview by Erika Nesvold, *Making New Worlds*, November 15, 2017, podcast audio and transcript, https://makingnewworlds.com/2017/11/15/episode-1-why-are-we-going/.

3. Darcie Little Badger, "Ep 01: Why Are We Going?," *Making New Worlds*, November 15, 2017, podcast audio and transcript, https://makingnewworlds.com/2017/11/15/episode-1-why-are-we-going/.

4. Robert Zubrin, *The Case for Mars: The Plan to Settle the Red Planet and Why We Must* (New York: Touchstone, 1996).

5. T. L. Mitchell, *Three Expeditions into the Interior of Eastern Australia*, vol. 1 (London: T. and W. Boone, 1839), 90.

6. Bruce Pascoe, *Dark Emu: Aboriginal Australia and the Birth of Agriculture* (London: Scribe, 2018).

7. Donna Gabaccia, "Ep 01: Why Are We Going?," interview by Erika Nesvold, *Making New Worlds*, November 15, 2017, podcast audio and transcript, https://makingnewworlds.com/2017/11/15/episode-1-why-are-we-going/.

8. Laurence Bergreen, *Over the Edge of the World: Magellan's Terrifying Circumnavigation of the Globe* (New York: Harper Collins, 2003).

9. Morgan Stanley, *Investment Implications of the Final Frontier*, October 12, 2017, http://www.fullertreacymoney.com/system/data/files/PDFs/2017/October/20th/msspace.pdf.

10. Gabaccia, "Ep 01: Why Are We Going?"

11. "Trans-Atlantic Slave Trade Database," Slave Voyages, accessed February 1, 2020, https://www.slavevoyages.org/assessment/estimates.

12. "Enumeration of Persons in the Several Districts of the United States," US Census Bureau, December 8, 1801, https://www2.census.gov/library/publications/decennial/1800/1800-returns.pdf.

13. Tom McKay, "Elon Musk: A New Life Awaits You in the Off-World Colonies—for a Price," *Gizmodo*, January 17, 2020, https://gizmodo.com/elon-musk-a-new-life-awaits-you-on-the-off-world-colon-1841071257.

14. Farley Grubb, "The Incidence of Servitude in Trans-Atlantic Migration, 1771–1804," *Explorations in Economic History* 22, no. 3 (1985): 316–339, https://doi.org/10.1016/0014-4983(85)90016-6.

15. Logan Marshall, ed., *On Board the Titanic: The Complete Story with Eyewitness Accounts* (Mineola, NY: Dover, 2006).

16. Garrett Hardin, "Lifeboat Ethics: The Case Against Helping the Poor," *Psychology Today Magazine* 8 (1974): 38–43, http://www.garretthardinsociety.org/articles/art_lifeboat_ethics_case_against_helping_poor.html; Garrett Hardin, "Living on a Lifeboat," *BioScience* 24, no. 10 (1974): 561–568, http://www.garretthardinsociety.org/articles/art_living_on_a_lifeboat.html.

17. Gabaccia, "Ep 01: Why Are We Going?"

CHAPTER 3

1. Kami White, "Virgin Galactic Welcomes Two New Pilots," Virgin Galactic, October 27, 2020, https://www.virgin.com/about-virgin/latest/virgin-galactic-welcomes-two-new-pilots.

2. NASA, "NASA's Newest Astronaut Recruits to Conduct Research off the Earth, for the Earth and Deep Space Missions," press release, June 7, 2017, https://www.nasa.gov/press-release/nasa-s-newest-astronaut-recruits-to-conduct-research-off-the-earth-for-the-earth-and.

3. UN General Assembly, "Universal Declaration of Humans Rights," 217 (III) A (Paris, 1948), http://www.un.org/en/universal-declaration-human-rights/.

4. UN General Assembly, "Treaty on Principles Governing the Activities of States in the Exploration and Use of Outer Space, Including the Moon and Other Celestial Bodies," 2222 (XXI) (1967), http://www.unoosa.org/oosa/en/ourwork/spacelaw/treaties/outerspacetreaty.html.

5. Douglas Adams, *The Hitchhiker's Guide to the Galaxy* (New York: Harmony Books, 1980).

6. David E. Sanger, "Soviets Send First Japanese, a Journalist, into Space," *New York Times*, December 3, 1990, https://nyti.ms/2O217Nh. In order, the seven citizens are: Dennis Tito, Mark Shuttleworth, Gregory Olsen, Anousheh Ansari, Charles Simonyi, Richard Garriott, and Guy Laliberté. Charles Simonyi has been to space twice. All eight flights were arranged by the space tourism company Space Adventures. More information on these flights can be found at the Space Adventures website: http://www.spaceadventures.com/experiences/space-station/.

7. Kenneth Chang, "A Billionaire Names His Team to Ride SpaceX, No Pros Allowed," *New York Times*, March 30, 2021, https://www.nytimes.com/2021/03/30/science/30spacex-inspiration4.html.

8. Sean Potter, ed., "Explorers Wanted: NASA to Hire More Artemis Generation Astronauts," NASA press release, February 11, 2020, https://www.nasa.gov/press-release/explorers-wanted-nasa-to-hire-more-artemis-generation-astronauts.

9. "List of Space Travelers by Name," Wikipedia, the Free Encyclopedia, accessed May 8, 2021, https://en.wikipedia.org/wiki/List_of_space_travelers_by_name.

10. "Population, Female (% of Total Population)," World Bank Group, accessed January 7, 2020, https://data.worldbank.org/indicator/sp.pop.totl.fe.zs.

11. "Top 10 Most Populous Countries (July 1, 2019)," US Census Bureau, accessed January 7, 2020, https://www.census.gov/popclock/print.php?component=counter.

12. Henry P. Stewart, "The Impact of the USS *Forrestal*'s 1967 Fire on United States Navy Shipboard Damage Control" (master's thesis, US Army Command and General Staff College, 2004), https://apps.dtic.mil/sti/pdfs/ADA429103.pdf; "Trial by Fire: A Carrier Fights For Life," Naval Photographic Center, produced 1973, educational video, 18:43, https://www.youtube.com/watch?v=U6NnfRT_OZA.

13. "Psychological and Medical Selection Process," European Space Agency, accessed April 3, 2021, https://www.esa.int/Science_Exploration/Human_and_Robotic_Exploration/European_Astronaut_Selection_2008/Psychological_and_medical_selection_process.

14. Matthew Cantor, "NASA Cancels All-Female Spacewalk, Citing Lack of Spacesuit in Right Size," *The Guardian*, March 26, 2019, https://www.theguardian.com/science/2019/mar/25/nasa-all-female-spacewalk-canceled-women-spacesuits.

15. Jesse Shanahan, "Ep 02: Who Gets to Go?," interview by Erika Nesvold, *Making New Worlds*, November 22, 2017, podcast audio and transcript, https://makingnewworlds.com/2017/11/22/episode-2-who-gets-to-go/.

16. "Parastronaut Feasibility Project," European Space Agency, accessed April 3, 2021, https://www.esa.int/About_Us/Careers_at_ESA/ESA_Astronaut_Selection/Parastronaut_feasibility_project.

17. Luca Parmitano, "EVA 23: Exploring the Frontier," *Luca blog*, European Space Agency, August 20, 2013, http://blogs.esa.int/luca-parmitano/2013/08/20/eva-23-exploring-the-frontier/.

18. Cameron M. Smith, "Estimation of a Genetically Viable Population for Multigenerational Interstellar Voyaging: Review and Data for Project Hyperion," *Acta Astronautica* 97 (2014): 16–29, https://doi.org/10.1016/j.actaastro.2013.12.013.

19. Elon Musk, "Making Humans a Multi-Planetary Species," *New Space* 5, no. 2 (2017): 46–61, https://doi.org/10.1089/space.2017.29009.emu.

20. Ivor Noël Hume, "We Are Starved," *Colonial Williamsburg Journal* 29, no. 1 (2007): 44–51, https://research.colonialwilliamsburg.org/Foundation/journal/Winter07/starving.cfm.

21. David A. Price, *Love and Hate in Jamestown: John Smith, Pocahontas, and the Start of a New Nation*, reprint ed. (New York: Vintage, 2007).

CHAPTER 4

1. Amanda Nguyen et al., "Law & Order or Game of Thrones? The Legal Landscape of Space Exploration," panel discussion, New America's How Will We Govern Ourselves in Space?, Washington, DC, July 10, 2019, http://opentranscripts.org/transcript/legal-landscape-space-exploration/.

2. Laura Montgomery, "Ep 03: Who Owns Mars?," interview by Erika Nesvold, *Making New Worlds*, November 29, 2017, podcast audio and transcript, https://makingnewworlds.com/2017/11/29/ep-03-who-owns-mars/.

3. Jeff Foust, "Bigelow to Press US Government on Lunar Property Rights," *Space Politics*, November 13, 2013, http://www.spacepolitics.com/2013/11/13/bigelow-to-press-us-government-on-lunar-property-rights/.

4. Garrett Hardin, "The Tragedy of the Commons," *Science* 162, no. 3859 (1968): 1243–1248, http://doi.org/10.1126/science.162.3859.1243.

5. Elinor Ostrom, *Governing the Commons: The Evolution of Institutions for Collective Actions* (Cambridge: Cambridge University Press, 1990).

6. Nguyen et al., "Law & Order, or Game of Thrones?"

7. Debbie Becher, "Ep 03: Who Owns Mars?," interview by Erika Nesvold, *Making New Worlds*, November 29, 2017, podcast audio and transcript, https://makingnewworlds.com/2017/11/29/ep-03-who-owns-mars/.

8. Becher, "Ep 03: Who Owns Mars?"

9. Becher, "Ep 03: Who Owns Mars?"

10. Henry R. Hertzfeld, Brian Weeden, and Christopher D. Johnson, "How Simple Terms Mislead Us: The Pitfalls of Thinking about Outer Space as a Commons," *International Astronautical Congress* 15, 2015, https://swfound.org/media/205390/how-simple-terms-mislead-us-hertzfeld-johnson-weeden-iac-2015.pdf.

11. Barbara Arneil, "The Wild Indian's Venison: Locke's Theory of Property and English Colonialism in America," *Political Studies* 44, no. 1 (1996): 60–74, https://journals.sagepub.com/doi/10.1111/j.1467-9248.1996.tb00764.x.

12. Kenneth H. Bobroff, "Retelling Allotment: Indian Property Rights and the Myth of Common Ownership," *Vanderbilt Law Review* 54, no. 4 (2001): 1559–1623, https://scholarship.law.vanderbilt.edu/cgi/viewcontent.cgi?article=1879&context=vlr.

13. Margaret Newell, "Ep 03: Who Owns Mars?," interview by Erika Nesvold, *Making New Worlds*, November 29, 2017, podcast audio and transcript, https://making newworlds.com/2017/11/29/ep-03-who-owns-mars/.

14. Newell, "Ep 03: Who Owns Mars?"

15. Melinda C. Miller, "Land and Racial Wealth Inequality," *American Economic Review: Papers & Proceedings* 101, no. 3 (2011): 371–376, https://doi.org/10.1257 /aer.101.3.371.

16. E. H. P. Frankema, "The Colonial Origins of Inequality: Exploring the Causes and Consequences of Land Distribution," *IAI Discussion Papers*, no. 119 (2005), https://www.econstor.eu/bitstream/10419/27410/1/504473565.PDF.

17. W. T. Sherman, *Special Field Orders, No. 15*, Headquarters Military Division of the Mississippi, January 16, 1865, in *The Wartime Genesis of Free Labor: The Lower South*, ed. Ira Berlin, Thavolia Glymph, Steven F. Miller, Joseph P. Reidy, Leslie S. Rowland, and Julie Saville (Cambridge, NY: Cambridge University Press, 2012), 338–340, http://www.freedmen.umd.edu/sfo15.htm.

18. Danielle Alexander, "Forty Acres and a Mule: The Ruined Hope of Reconstruction," *Humanities* 25, no. 1 (2004), https://web.archive.org/web/20080916095443 /http://neh.gov/news/humanities/2004-01/reconstruction.html.

19. UN General Assembly, "Treaty on Principles Governing the Activities of States in the Exploration and Use of Outer Space, Including the Moon and Other Celestial Bodies," 2222 (XXI) (1967), http://www.unoosa.org/oosa/en/ourwork/spacelaw /treaties/outerspacetreaty.html.

20. Tim Fernholz, "The US Government Has Approved the First Private Landing on the Moon," *Quartz*, August 3, 2016, https://qz.com/749246/the-us-govern ment-has-approved-the-first-private-landing-on-the-moon/.

21. "69 of the Richest 100 Entities on the Planet Are Corporations, Not Governments, Figures Show," Global Justice Now, October 17, 2018, https://www .globaljustice.org.uk/news/2018/oct/17/69-richest-100-entities-planet-are-corpo rations-not-governments-figures-show.

22. Montgomery, "Ep 03: Who Owns Mars?"

23. UN General Assembly, "Agreement Governing the Activities of States on the Moon and Other Celestial Bodies," RES 34/68 (1979), https://www.unoosa.org /oosa/en/ourwork/spacelaw/treaties/moon-agreement.html.

24. Jeff Foust, "New Law Unlikely to Settle Debate on Space Resource Rights," *SpaceNews*, December 4, 2015, https://spacenews.com/new-law-unlikely-to-settle-debate -on-space-resource-rights/.

25. US Commercial Space Launch Competitiveness Act, H.R. 2262, 114th Cong., 2015, https://www.congress.gov/bill/114th-congress/house-bill/2262.

26. Fabio Tronchetti, "The Space Resource Exploration and Utilization Act: A Move Forward or a Step Back?," *Space Policy* 34 (2015): 6–10, https://doi.org/10.1016/j .spacepol.2015.08.001.

27. See, for example, Leviticus 25:22, Numbers 27:8–11, and 1 Kings 21:2–3 (all NIV).

28. John P. Powelson, *The Story of Land: A World History of Land Tenure and Agrarian Reform* (Cambridge, MA: Lincoln Institute of Land Policy, 1988).

29. Karl Marx and Frederick (Friedrich) Engels, *Manifesto of the Communist Party*, 1848, published in *Marx/Engels Selected Works, Vol. One* (Moscow: Progress Publishers, 1969), 22, https://www.marxists.org/archive/marx/works/download/pdf/Manifesto.pdf.

30. Newell, "Ep 03: Who Owns Mars?"

31. Newell, "Ep 03: Who Owns Mars?"

32. Julian Brave NoiseCat, interview with the author, June 25, 2020.

33. Bobroff, "Retelling Allotment," 1559–1623.

34. Charles C. Mann, *1491: New Revelations of the Americas Before Columbus*, 2nd ed. (New York: Vintage, 2006), 302.

35. Jeremy Lurgio, "Saving the Whanganui: Can Personhood Rescue a River?," *The Guardian*, November 29, 2019, https://www.theguardian.com/world/2019/nov/30/saving-the-whanganui-can-personhood-rescue-a-river.

36. Becher, "Ep 03: Who Owns Mars?"

37. Alice Gorman, "Can the Moon Be a Person? As Lunar Mining Looms, a Change of Perspective Could Protect Earth's Ancient Companion," *The Conversation*, August 26, 2020, https://theconversation.com/can-the-moon-be-a-person-as-lunar-mining-looms-a-change-of-perspective-could-protect-earths-ancient-companion-144848.

38. UN General Assembly, "Treaty on Principles Governing the Activities of States in the Exploration and Use of Outer Space."

CHAPTER 5

1. S. Nazrul Islam and John Winkel, "Climate Change and Social Inequality," DESA Working Paper No. 152 ST/ESA/2017/DWP/152, United Nations Department of Economic & Social Affairs, 2017, https://www.un.org/esa/desa/papers/2017/wp152_2017.pdf.

2. Michael L. Ross, "What Have We Learned about the Resource Curse?," *Annual Review of Political Science* 18 (2015): 239–259, https://doi.org/10.1146/annurev-polisci-052213-040359.

3. Michael L. Ross, *The Oil Curse: How Petroleum Wealth Shapes the Development of Nations* (Princeton, NJ: Princeton University Press, 2012).

4. "Space Debris by the Numbers," European Space Agency, accessed November 1, 2020, https://www.esa.int/Safety_Security/Space_Debris/Space_debris_by_the_numbers.

5. "Space Debris by the Numbers," European Space Agency.

6. "Frequently Asked Questions," NASA Orbital Debris Program Office, accessed November 1, 2020, https://orbitaldebris.jsc.nasa.gov/faq/.

7. UN General Assembly, "Treaty on Principles Governing the Activities of States in the Exploration and Use of Outer Space, Including the Moon and Other Celestial Bodies," 2222 (XXI) (1967), http://www.unoosa.org/oosa/en/ourwork /spacelaw/treaties/outerspacetreaty.html.

8. Donald J. Kessler and Burton G. Cour-Palais, "Collision Frequency of Artificial Satellites: The Creation of a Debris Belt," *Journal of Geophysical Research* 83, no. A6 (1978): 2637–2646, https://doi.org/10.1029/JA083iA06p02637.

9. Moriba Jah, interview with the author, October 26, 2020.

10. Francesca Letizia et al., "Application of a Debris Index for Global Evaluation of Mitigation Strategies," *Acta Astronautica* 161 (2019): 348–362, https://doi.org /10.1016/j.actaastro.2019.05.003.

11. Jah, interview.

12. "ESA Commissions World's First Space Debris Removal," European Space Agency, accessed September 12, 2019, https://www.esa.int/Safety_Security/Clean _Space/ESA_commissions_world_s_first_space_debris_removal; Mike Wall, "Foam 'Spider Webs' from Tiny Satellites Could Help Clean Up Space Junk," *Space.com*, June 23, 2020, https://www.space.com/space-junk-cleanup-foam-satellite-tech nology.html.

13. Chris Newman, "Ep 08: Should We Make Mars More Like Earth?," interview by Erika Nesvold, *Making New Worlds*, January 17, 2018, podcast audio and transcript, https://makingnewworlds.com/2018/01/17/ep-08-should-we-make-mars-more -like-earth-terraforming-and-environmental-conservation/.

14. Declaration of the First Meeting of Equatorial Countries, Bra.-Col.-Cog.-Ecu.- Idn.-Ken.-Uga.-Zar., December 3, 1976, https://www.jaxa.jp/library/space_law /chapter_2/2-2-1-2_e.html.

15. Scott Ervin, "Law in a Vacuum: The Common Heritage Doctrine in Outer Space Law," *Boston College International and Comparative Law Review* 7, no. 2 (1984): 403–431, http://lawdigitalcommons.bc.edu/iclr/vol7/iss2/9.

16. UN General Assembly, "Treaty on Principles Governing the Activities of States in the Exploration and Use of Outer Space."

17. James S. J. Schwartz, "Ep 08: Should We Make Mars More Like Earth?," interview by Erika Nesvold, *Making New Worlds*, January 17, 2018, podcast audio and transcript, https://makingnewworlds.com/2018/01/17/ep-08-should-we-make-mars -more-like-earth-terraforming-and-environmental-conservation/.

18. Rachel Riederer, "Silicon Valley Says Space Mining Is Awesome and Will Change Life on Earth. That's Only Half Right," *New Republic*, May 19, 2014, https://new republic.com/article/117815/space-mining-will-not-solve-earths-conflict-over -natural-resources.

19. Martin Elvis, "How Many Ore-Bearing Asteroids?," *Planetary and Space Science* 91 (2014): 20–26, http://doi.org/10.1016/j.pss.2013.11.008.

20. Martin Elvis, Alanna Krolikowski, and Tony Milligan, "Concentrated Lunar Resources: Imminent Implications for Governance and Justice," *Philosophical Transactions of the Royal Society A* 379, no. 2188 (2020), http://doi.org/10.1098/rsta.2019.0563.

21. Martin Elvis, Tony Milligan, and Alanna Krolikowski, "The Peaks of Eternal Light: A Near-Term Property Issue on the Moon," *Space Policy* 38 (2016): 30–38, http://doi.org/10.1016/j.spacepol.2016.05.011.

22. Martin Elvis, interview with the author, October 19, 2020.

23. UN General Assembly, "Treaty on Principles Governing the Activities of States in the Exploration and Use of Outer Space."

24. Schwartz, "Ep 08: Should We Make Mars More Like Earth?"

25. UN General Assembly, "Treaty on Principles Governing the Activities of States in the Exploration and Use of Outer Space."

26. Elvis, interview.

27. Al Gore, *Earth in the Balance: Ecology and the Human Spirit* (New York: Houghton Mifflin, 1992), 170.

28. Roman Krznaric, *The Good Ancestor: A Radical Prescription for Long-Term Thinking* (New York: The Experiment, 2020), 72.

29. Oren Lyons, "An Iroquois Perspective," in *American Indian Environments: Ecological Issues in Native American History*, ed. Christopher Vecsey and Robert W. Venables (Syracuse, NY: Syracuse University Press, 1980), 172.

30. John Rawls, *A Theory of Justice* (Cambridge, MA: Belknap Press, 1971).

31. Edith Brown Weiss, "In Fairness to Future Generations and Sustainable Development," *American University International Law Review* 8, no. 1 (1992): 19–26, https://digitalcommons.wcl.american.edu/auilr/vol8/iss1/2/.

32. Edith Brown Weiss, "In Fairness to Future Generations," *Environment: Science and Policy for Sustainable Development* 32, no. 3 (1990): 6–31, https://doi.org/10.1080/00139157.1990.9929015.

33. Alice Vincent, "Ombudspersons for Future Generations: Bringing Intergenerational Justice into the Heart of Policymaking," *UN Chronicle* 49, no. 2 (2012): 66–68, https://doi.org/10.18356/2c3c1e22-en.

34. Kate Raworth, "A Safe and Just Space for Humanity," *Oxfam Discussion Papers* (2012), https://www.oxfam.org/en/research/safe-and-just-space-humanity.

35. Kyle Whyte, interview with the author, March 29, 2021.

36. Eric Holthaus, *The Future Earth: A Radical Vision for What's Possible in the Age of Warming* (New York: HarperCollins, 2020), 119.

37. UN General Assembly, "Treaty on Principles Governing the Activities of States in the Exploration and Use of Outer Space."

38. Charles Cockell and Gerda Horneck, "Planetary Parks—Formulating a Wilderness Policy for Planetary Bodies," *Space Policy* 22, no. 4 (2006): 256–261, https://doi.org/10.1016/j.spacepol.2006.08.006.

39. Charles Cockell and Gerda Horneck, "A Planetary Park System for Mars," *Space Policy* 20, no. 4 (2004): 291–295, https://doi.org/10.1016/j.spacepol.2004.08.003.

40. Daniel Capper, "Preserving Mars Today Using Baseline Ecologies," *Space Policy* 49 (2019), https://doi.org/10.1016/j.spacepol.2019.05.003.

41. Schwartz, "Ep 08: Should We Make Mars More Like Earth?"

42. Jah, interview.

43. Martin Elvis and Tony Milligan, "How Much of the Solar System Should We Leave as Wilderness?," *Acta Astronautica* 162 (2019): 574–580, https://doi.org/10.48550/arXiv.1905.13681.

44. Elvis, interview.

CHAPTER 6

1. Aldo Leopold, "The Land Ethic," in *A Sand County Almanac: And Sketches Here and There* (Oxford: Oxford University Press, 1949), 224–225.

2. Ronald Wright, *A Short History of Progress* (Toronto: House of Anansi, 2004), 37.

3. Enrique H. Bucher, "The Causes of Extinction of the Passenger Pigeon," in *Current Ornithology*, vol. 9, ed. Dennis M. Power (Boston: Springer, 1992), 1–36, https://link.springer.com/chapter/10.1007%2F978-1-4757-9921-7_1; Jessica E. Thomas et al., "Demographic Reconstruction from Ancient DNA Supports Rapid Extinction of the Great Auk," *eLife* 8 (2019): e47509, https://doi.org/10.7554%2FeLife.47509.

4. UN Food and Agriculture Organization, *The State of World Fisheries and Aquaculture 2020: Sustainability in Action*, Rome: FAO, 2020, https://doi.org/10.4060/ca9229en.

5. David D. Smits, "The Frontier Army and the Destruction of the Buffalo: 1865–1883," *Western Historical Quarterly* 25, no. 3 (1994): 312–338, https://doi.org/10.2307/971110.

6. David S. Wilcove et al., "Quantifying Threats to Imperiled Species in the United States," *BioScience* 48, no. 8 (1998): 607–615, https://doi.org/10.2307/1313420.

7. Robin McKie, "How Our Colonial Past Altered the Ecobalance of an Entire Planet," *The Guardian*, June 10, 2018, https://www.theguardian.com/science/2018/jun/10/colonialism-changed-earth-geology-claim-scientists.

8. Brian D. Cooke, "Rabbits: Manageable Environmental Pests of Participants in New Australian Ecosystems?," *Wildlife Research* 39, no. 4 (2012): 279–289, http://doi.org/10.1071/WR11166.

9. UN General Assembly, "Treaty on Principles Governing the Activities of States in the Exploration and Use of Outer Space, Including the Moon and Other Celestial Bodies," 2222 (XXI) (1967), http://www.unoosa.org/oosa/en/ourwork/spacelaw/treaties/outerspacetreaty.html.

10. William M. Denevan, "The Pristine Myth: The Landscape of the Americas in 1492," *Annals of the Association of American Geographers* 82, no. 3 (1992): 369–385, https://doi.org/10.1111/j.1467-8306.1992.tb01965.x.

11. Meghan Bartels, "Apollo 11 Astronauts Spent 3 Weeks in Quarantine, Just in Case of Moon Plague," *Space.com*, July 24, 2019, https://www.space.com/apollo-11-astronauts-quarantined-after-splashdown.html.

12. Margaret Race, "Ep 09: What If There's Already Life on Mars?," interview by Erika Nesvold, *Making New Worlds*, January 24, 2018, podcast audio and transcript, https://makingnewworlds.com/2018/01/24/ep-09-what-if-theres-already-life-on-mars-planetary-protection/.

13. Kelly Smith, "Ep 09: What If There's Already Life on Mars?," interview by Erika Nesvold, *Making New Worlds*, January 24, 2018, podcast audio and transcript, https://makingnewworlds.com/2018/01/24/ep-09-what-if-theres-already-life-on-mars-planetary-protection/.

14. "COSPAR Policy on Planetary Protection," COSPAR Panel on Planetary Protection, June 17, 2020, https://cosparhq.cnes.fr/assets/uploads/2020/07/PPPolicy June-2020_Final_Web.pdf.

15. Race, "Ep 09: What If There's Already Life on Mars?"

16. Smith, "Ep 09: What If There's Already Life on Mars?"

17. Mark Lupisella and John Logsdon, "Do We Need a Cosmocentric Ethic?" (paper IAA-97-IAA.9.2.09, International Astronautical Federation Congress, Turin, Italy, 1997), https://www.academia.edu/266597/Do_We_Need_a_Cosmocentric_Ethic.

18. Daniel Oberhaus, "A Crashed Israeli Lunar Lander Spilled Tardigrades on the Moon," *Wired*, August 5, 2019, https://www.wired.com/story/a-crashed-israeli-lunar-lander-spilled-tardigrades-on-the-moon/.

19. Chris Taylor, "'I'm the First Space Pirate!' How Tardigrades Were Secretly Smuggled to the Moon," *Mashable*, August 8, 2019, https://mashable.com/article/smuggled-moon-tardigrade/.

20. Robert Zubrin, "The Tardigrades-on-the-Moon Affair," *National Review*, August 31, 2019, https://www.nationalreview.com/2019/08/planetary-protection-rules-hamper-space-exploration/.

21. Christopher P. McKay, "On Terraforming Mars," *Extrapolation* 23, no. 4 (1982): 309–314, https://doi.org/10.3828%2Fextr.1982.23.4.309.

22. Carl Sagan, "The Planet Venus," *Science* 133, no. 3456 (1961): 849–858, https://doi.org/10.1126%2Fscience.133.3456.849.

23. Carl Sagan, "Planetary Engineering on Mars," *Icarus* 20, no. 4 (1973): 513–514, https://doi.org/10.1016%2F0019-1035%2873%2990026-2.

24. Christopher P. McKay, Owen B. Toon, and James F. Kasting, "Making Mars Habitable," *Nature* 352 (1991): 489–496, https://doi.org/10.1038/352489a0; Robert M. Zubrin and Christopher P. McKay, "Technological Requirements for Terraforming Mars," *AIAA-03–2005* (1993), https://doi.org/10.2514/6.1993-2005; J. L. Green et al., "A Future Mars Environment for Science and Exploration" (paper presented at the Planetary Science Vision 2050 Workshop, Washington, DC, February 2017), https://www.hou.usra.edu/meetings/V2050/pdf/8250.pdf; Julian A. Hiscox and David J. Thomas, "Genetic Modification and Selection of Microorganisms

for Growth on Mars," *Journal of the British Interplanetary Society* 48, no. 10 (1995): 419–426, https://pubmed.ncbi.nlm.nih.gov/11541203/.

25. Robert Zubrin, *The Case for Mars: The Plan to Settle the Red Planet and Why We Must* (New York: Touchstone, 1996).

26. Carl Sagan, *Cosmos* (New York: Random House, 1980), 138.

27. Christopher P. McKay, "Does Mars Have Rights? An Approach to the Environmental Ethics of Planetary Engineering," in *Moral Expertise: Studies in Practical and Professional Ethics*, ed. Don MacNiven (New York: Routledge, 1990).

28. Smith, "Ep 09: What If There's Already Life on Mars?"

29. Bernard Dixon, "Smallpox—Imminent Extinction, and an Unresolved Dilemma," *New Scientist* 69, no. 989 (1976): 430–432.

30. James Schwartz, "Ep 08: Should We Make Mars More Like Earth?," interview by Erika Nesvold, *Making New Worlds*, January 17, 2018, podcast audio and transcript, https://makingnewworlds.com/2018/01/17/ep-08-should-we-make-mars-more-like-earth-terraforming-and-environmental-conservation/.

31. Wilderness Act of 1964, Pub. L. 88-577, 16 U.S.C. 1131-1136, https://www.congress.gov/88/statute/STATUTE-78/STATUTE-78-Pg890.pdf.

32. Holmes Rolston III, "The Preservation of Natural Value in the Solar System," in *Beyond Spaceship Earth: Environmental Ethics and the Solar System*, ed. Eugene C. Hargrove (San Francisco: Sierra Club Books, 1986), 178.

33. Martyn J. Fogg, "The Ethical Dimensions of Space Settlement," *Space Policy* 16, no. 3 (2000): 205–211, http://doi.org/10.1016/S0265-9646(00)00024-2.

34. Robert Sparrow, "The Ethics of Terraforming," *Environmental Ethics* 21, no. 3 (1999): 227–245, https://doi.org/10.5840/enviroethics199921315.

35. Schwartz, "Ep 08: Should We Make Mars More Like Earth?"

36. Smith, "Ep 09: What If There's Already Life on Mars?"

37. Robert Heath French, "Environmental Philosophy and the Ethics of Terraforming Mars: Adding the Voices of Environmental Justice and Ecofeminism to the Ongoing Debate" (master's thesis, University of North Texas, August 2013).

CHAPTER 7

1. Edgar Zapata, "An Assessment of Cost Improvements in the NASA COTS-CSR Program and Implications for Future NASA Missions," NASA (2018), https://ntrs.nasa.gov/archive/nasa/casi.ntrs.nasa.gov/20170008895.pdf.

2. Will Wei, "Peter Diamandis: The First Trillionaire Is Going to Be Made in Space," *Business Insider*, March 2, 2015, https://www.businessinsider.com/peter-diamandis-space-trillionaire-entrepreneur-2015-2; Katie Kramer, "Neil deGrasse Tyson Says Space Ventures Will Spawn First Trillionaire," *NBC News*, May 3, 2015, https://www.nbcnews.com/science/space/neil-degrasse-tyson-says-space-ventures-will-spawn-first-trillionaire n352271.

3. Christian Davenport, "Elon Musk on Mariachi Bands, Zero-G Games, and Why His Mars Plan Is Like 'Battlestar Galactica,'" *Washington Post*, September 28, 2016, https://www.washingtonpost.com/news/the-switch/wp/2016/09/28/elon-musk-on-mariachi-bands-how-he-plans-to-make-space-travel-fun-and-why-his-mars-plan-is-like-battlestar-galactica/.

4. Nicole Dieker, "Mars: The Rich Planet," *Medium*, September 28, 2018, https://medium.com/the-billfold/mars-the-rich-planet-e6be98360f2b.

5. Matthew Weinzierl, "Space, the Final Economic Frontier," *Journal of Economic Perspectives* 32, no. 2 (2018): 173–192, https://www.jstor.org/stable/26409430.

6. Margaret Newell, "Ep 04: Where's My Money?," interview by Erika Nesvold, *Making New Worlds*, December 6, 2017, podcast audio and transcript, https://makingnewworlds.com/2017/12/06/ep-04-wheres-my-money/.

7. Susanne Barton and Hannah Recht, "The Massive Prize Luring Miners to the Stars," *Bloomberg*, March 8, 2018, https://www.bloomberg.com/graphics/2018-asteroid-mining/.

8. Margaret Newell, "Ep 04: Where's My Money?"

9. Adam Thiere, *Permissionless Innovation: The Continuing Case for Comprehensive Technological Freedom* (Arlington, VA: Mercatus Center at George Mason University, 2016).

10. Eli Dourado, *Regulating Space: Innovation, Liberty, and International Obligations: Hearing before the Science, Space, and Technology Subcommittee on Space, House of Representatives*, 115th Cong. (March 8, 2017), https://www.mercatus.org/publications/technology-and-innovation/creating-environment-permissionless-innovation-outer-space.

11. Robert Zubrin, *The Case for Mars: The Plan to Settle the Red Planet and Why We Must* (New York: Touchstone, 1996).

12. Margaret Newell, "Ep 04: Where's My Money?"

13. Zubrin, *The Case for Mars*.

14. Roger D. Launius, "The Railroads and the Space Program Revisited: Historical Analogues and the Stimulation of Commercial Space Operations," *Astropolitics* 12, no. 2–3 (2014): 167–179, https://doi.org/10.1080/14777622.2014.964129; Adi Robertson, "SpaceX Wants to Be the Railroad of the Future," *The Verge*, September 27, 2016, https://www.theverge.com/2016/9/27/13080970/spacex-elon-musk-mars-expedition-railroad-of-the-future.

15. Gordon H. Chang, Shelley Fisher Fishkin, and Hilton Odenzinger, intro to *The Chinese and the Iron Road: Building the Transcontinental Railroad*, ed. Gordon H. Chang and Shelley Fisher Fishkin (Stanford: Stanford University Press, 2019).

16. Sarah Newell, "Ep 04: Where's My Money?," interview by Erika Nesvold, *Making New Worlds*, December 6, 2017, podcast audio and transcript, https://makingnewworlds.com/2017/12/06/ep-04-wheres-my-money/.

17. Karl Marx, *Critique of the Gotha Programme*, 1875, in *Marx/Engels Selected Works, Vol. Three* (Moscow: Progress Publishers, 1970), https://www.marxists.org/archive/marx/works/1875/gotha/ch01.htm.

18. Ester Bloom, "Elon Musk's SpaceX Shortchanged Its Workers and Now It Must Pay," *CNBC*, May 15, 2017, https://www.cnbc.com/2017/05/15/elon-musks-spacex -mistreated-its-workers-and-now-it-must-pay.html.

19. Loren Grush, "Jeff Bezos' Space Company Is Pressuring Employees to Launch a Tourist Rocket during the Pandemic," *The Verge*, April 2, 2020, https://www .theverge.com/2020/4/2/21198272/blue-origin-coronavirus-leaked-audio-test-launch -workers-jeff-bezos.

20. Julia Carrie Wong, "Tesla Factory Workers Reveal Pain, Injury and Stress: 'Everything Feels like the Future but Us,'" *The Guardian*, May 18, 2017, https://www .theguardian.com/technology/2017/may/18/tesla-workers-factory-conditions-elon -musk.

21. Kanishka Singh, "US Labor Judge Rules that Tesla Broke Labor Law," *Reuters*, September 27, 2019, https://www.reuters.com/article/us-tesla-labor/u-s-labor -judge-rules-that-tesla-broke-labor-law-idUSKBN1WD003.

22. Bryan Menegus, "Exclusive: Amazon's Own Numbers Reveal Staggering Injury Rates at Staten Island Warehouse," *Gizmodo*, November 25, 2019, https://gizmodo .com/exclusive-amazons-own-numbers-reveal-staggering-injury-1840025032.

CHAPTER 8

1. Will Stewart, "Antarctic Scientist 'Stabs Colleague Who Kept Telling Him the Endings of Books He Was Reading on Remote Research Station,'" *The Sun*, October 30, 2018, https://www.thesun.co.uk/news/7615571/antarctic-scientist-stabs -colleague-who-kept-telling-him-endings-of-books-he-was-reading/; Caroline Haskins, "An Attempted Murder at a Research Station Shows How Crimes Are Prosecuted in Antarctica," *Vice*, October 25, 2018, https://www.vice.com/en_us/article /xw9bg3/an-attempted-murder-at-a-research-station-shows-how-crimes-are-pros ecuted-in-antarctica.

2. Chris Newman, "Ep 05: What If Someone Steals My Stuff?," interview by Erika Nesvold, *Making New Worlds*, December 13, 2017, podcast audio and transcript, https://makingnewworlds.com/2017/12/13/ep-05-what-if-someone-steals-my-stuff/.

3. Scientific American Editors, "How to Reinvent Policing," *Scientific American* 323, no. 3, 8 (2020), https://www.scientificamerican.com/article/three-ways-to-fix-toxic -policing/.

4. Alex S. Vitale, *The End of Policing* (London: Verso, 2017).

5. Daniel S. Nagin, "Deterrence in the Twenty-First Century," *Crime and Justice* 42 (2013): 199–263, https://doi.org/10.1086/670398.

6. Justin Allen, interview with the author, October 24, 2020.

7. Charles S. Cockell, "Exoconfac—the Extraterrestrial Containment Facility: An Essay on the Design Philosophy for a Prison to Contain Criminals in Settlements Beyond the Earth," *Journal of the British Interplanetary Society* 69 (2016): 27–30, https://www.researchgate.net/publication/306208635_Exoconfac_-_The

_extraterrestrial_containment_facility_An_essay_on_the_design_philosophy
_for_a_prison_to_contain_criminals_in_settlements_beyond_the_earth.

8. Courtney Montoya, interview with the author, October 24, 2020.

9. Newman, "Ep 05: What If Someone Steals My Stuff?"

10. Patricia Leigh Brown, "Prison Makes Way for Future, but Preserves Past," *San Quentin Journal*, January 18, 2008, https://www.nytimes.com/2008/01/18/us/18 dungeon.html.

11. William Prochnau and Laura Parker, "Trouble in Paradise," *Vanity Fair*, December 17, 2007, https://www.vanityfair.com/news/2008/01/pitcairn200801.

12. Cockell, "Exoconfac."

13. Casey Williams, "The State Locked Them Up. Now They're North Carolina's Elite Firefighting Force," *Scalawag Magazine*, May 24, 2018, https://www.scala wagmagazine.org/2018/05/the-state-locked-them-up-now-theyre-north-caroli nas-elite-firefighting-force/.

14. Vanessa Romo, "California Bill Clears Path for Ex-Inmates to Become Fire-fighters," *NPR*, September 11, 2020, https://www.npr.org/2020/09/11/912193742 /california-bill-clears-path-for-ex-inmates-to-become-firefighters.

15. Jaime Lowe, "The Incarcerated Women Who Fight California's Wildfires," *New York Times Magazine*, August 31, 2017, https://www.nytimes.com/2017/08/31 /magazine/the-incarcerated-women-who-fight-californias-wildfires.html.

16. Otto Friedrich, *The Kingdom of Auschwitz* (New York: HarperCollins, 1994), 2.

17. AZ Rev Stat § 31–251 (2021), https://law.justia.com/codes/arizona/2021/title-31 /section-31-251/.

18. "About ACI," Arizona Correctional Industries, June 27, 2021, Wayback Machine, https://web.archive.org/web/20210627235219/https://aci.az.gov/about/.

19. "About ACI," Arizona Correctional Industries, February 17, 2022, Wayback Machine, https://web.archive.org/web/20220217021127/https://aci.az.gov/about/.

20. Lauren-Brooke Eisen, "Covid-19 Highlights the Need for Prison Labor Reform," *Brennan Center for Justice*, April 17, 2020, https://www.brennancenter.org /our-work/analysis-opinion/covid-19-highlights-need-prison-labor-reform.

21. US Constitution, amend. 13, sec. 1.

22. Douglas A. Blackmon, *Slavery by Another Name: The Re-Enslavement of Black Americans from the Civil War to World War II* (New York: Doubleday, 2008), 4.

23. Ed Pilkington, "US Inmates Stage Nationwide Prison Labor Strike over 'Modern Slavery,'" *The Guardian*, August 21, 2018, https://www.theguardian.com/us -news/2018/aug/20/prison-labor-protest-america-jailhouse-lawyers-speak.

24. Lowe, "The Incarcerated Women Who Fight California's Wildfires."

25. Michelle Brown, "Ep 05: What If Someone Steals My Stuff?," interview by Erika Nesvold, *Making New Worlds*, December 13, 2017, podcast audio and transcript, https://makingnewworlds.com/2017/12/13/ep-05-what-if-someone-steals-my-stuff/.

26. Michelle Alexander, *The New Jim Crow: Mass Incarceration in the Age of Colorblindness* (New York: New Press, 2010), 2.

27. *The Expanse*, season 4, episode 10, "Cibola Burn," directed by Breck Eisner, aired December 13, 2019, on Amazon Prime; *Battlestar Galactica*, season 1, episode 8, "Flesh and Bone," directed by Brad Turner, aired February 25, 2005, on Sci Fi.

28. Tom Godwin, "The Cold Equations," *Astounding Science Fiction* (August 1954), reprinted in *Lightspeed Magazine* 14 (July 2011), https://www.lightspeedmagazine.com/fiction/the-cold-equations/.

29. Ken Gonzales-Day, *Lynching in the West: 1850–1935* (Durham, NC: Duke University Press, 2006), 10.

30. Nick Herbert, "The Abolitionists' Criminal Conspiracy," *The Guardian*, July 27, 2008, https://www.theguardian.com/commentisfree/2008/jul/27/prisonsandprobation.youthjustice.

31. Brown, "Ep 05: What If Someone Steals My Stuff?"

32. Walidah Imarisha, "Ep 05: What If Someone Steals My Stuff?," interview by Erika Nesvold, *Making New Worlds*, December 13, 2017, podcast audio and transcript, https://makingnewworlds.com/2017/12/13/ep-05-what-if-someone-steals-my-stuff/.

33. Brown, "Ep 05: What If Someone Steals My Stuff?"

34. Heather Strang et al., "Restorative Justice Conferencing (RJC) Using Face-to-Face Meetings of Offenders and Victims: Effects on Offender Recidivism and Victim Satisfaction. A Systematic Review," *Campbell Systematic Reviews* 9, no. 1 (2013), https://doi.org/10.4073/csr.2013.12.

35. Imarisha, "Ep 05: What If Someone Steals My Stuff?"

36. Imarisha, "Ep 05: What If Someone Steals My Stuff?"

37. Brown, "Ep 05: What If Someone Steals My Stuff?"

CHAPTER 9

1. Mukesh Chiman Bhatt, "Space for Dissent: Disobedience on Artificial Habitats and Planetary Settlements," in *Dissent, Revolution and Liberty Beyond Earth*, ed. Charles S. Cockell (Cham, Switzerland: Springer, 2016), 78–92.

2. Michael Hiltzik, "The Day When Three NASA Astronauts Staged a Strike in Space," *Los Angeles Times*, December 28, 2015, https://www.latimes.com/business/hiltzik/la-fi-mh-that-day-three-nasa-astronauts-20151228-column.html.

3. David Shayler, *Around the World in 84 Days: The Authorized Biography of Skylab Astronaut Jerry Carr* (Burlington, ON: Apogee, 2008).

4. Peter Ward, *The Consequential Frontier: Challenging the Privatization of Space* (Brooklyn: Melville House, 2019), 166.

5. David Baker, "Liberty, Freedom and Democracy: Paradox for an Extraterrestrial Society," in *The Meaning of Liberty Beyond Earth*, ed. Charles S. Cockell (Cham, Switzerland: Springer, 2015), 247.

6. Tony Milligan, "Rawlsian Deliberation about Space Settlement," in *Human Governance Beyond Earth: Implications for Freedom*, ed. Charles S. Cockell (Cham, Switzerland: Springer, 2015), 10.

7. Tony Milligan, "Constrained Dissent and the Rights of Future Generations," in *Dissent, Revolution and Liberty Beyond Earth*, ed. Charles S. Cockell (Cham, Switzerland: Springer, 2016), 11; James S. J. Schwartz, "Fairness as a Moral Grounding for Space Policy," in *The Meaning of Liberty Beyond Earth*, ed. Charles S. Cockell (Cham, Switzerland: Springer, 2015), 69–89; John Rawls, *A Theory of Justice* (Cambridge, MA: Belknap Press, 1971).

8. Charles S. Cockell, "Freedom in a Box: Paradoxes in the Structure of Extraterrestrial Liberty," in *The Meaning of Liberty Beyond Earth*, ed. Charles S. Cockell (Cham, Switzerland: Springer, 2015), 58.

9. Charles S. Cockell, "Disobedience in Outer Space," in *Dissent, Revolution and Liberty Beyond Earth*, ed. Charles S. Cockell (Cham, Switzerland: Springer, 2016), 21.

10. Tony Milligan, "Constrained Dissent and the Rights of Future Generations," in *Dissent, Revolution and Liberty Beyond Earth*, ed. Charles S. Cockell (Cham, Switzerland: Springer, 2016), 19.

11. UN General Assembly, "Treaty on Principles Governing the Activities of States in the Exploration and Use of Outer Space, Including the Moon and Other Celestial Bodies," 2222 (XXI) (1967), http://www.unoosa.org/oosa/en/ourwork/spacelaw/treaties/outerspacetreaty.html.

12. Laura Montgomery, "Ep 06: Who's in Charge?," interview by Erika Nesvold, *Making New Worlds*, December 20, 2017, podcast audio and transcript, https://makingnewworlds.com/2017/12/20/ep-06-whos-in-charge/.

13. Laura Montgomery, *Regulating Space: Innovation, Liberty, and International Obligations: Hearing before the Subcommittee on Space, Committee on Science, Space, and Technology, House of Representatives*, 115th Cong. (March 8, 2017), https://www.govinfo.gov/content/pkg/CHRG-115hhrg24671/html/CHRG-115hhrg24671.htm.

14. Peter Ward, *The Consequential Frontier: Challenging the Privatization of Space* (Brooklyn: Melville House, 2019), 196.

15. Loren Grush, "NASA Announces International Artemis Accords to Standardize How to Explore the Moon," *The Verge*, May 15, 2020, https://www.theverge.com/2020/5/15/21259946/nasa-artemis-accords-lunar-exploration-moon-outer-space-treaty.

16. Peter Ward, *The Consequential Frontier*, 200.

17. Erik Persson, "Citizens of Mars Ltd.," in *Human Governance Beyond Earth: Implications for Freedom*, ed. Charles S. Cockell (Cham, Switzerland: Springer, 2015), 123.

18. "Starlink Pre-Order Agreement," Starlink, accessed April 7, 2021, https://www
    .starlink.com/legal/documents/DOC-1020-91087-64?regionCode=US.

19. Antonino Salmeri, "No, Mars Is Not a Free Planet, No Matter What SpaceX
    Says," *SpaceNews*, December 5, 2020, https://spacenews.com/op-ed-no-mars-is
    -not-a-free-planet-no-matter-what-spacex-says/.

20. Lauren Benton, "Ep 06: Who's in Charge?," interview by Erika Nesvold, *Making
    New Worlds*, December 20, 2017, podcast audio and transcript, https://making
    newworlds.com/2017/12/20/ep-06-whos-in-charge/.

21. Stephen Baxter, "The Cold Equations: Extraterrestrial Liberty in Science Fic-
    tion," in *The Meaning of Liberty Beyond Earth*, ed. Charles S. Cockell (Cham, Swit-
    zerland: Springer, 2015), 20.

22. Benton, "Ep 06: Who's in Charge?"

23. Montgomery, "Ep 06: Who's in Charge?"

24. Benton, "Ep 06: Who's in Charge?"

25. Benton, "Ep 06: Who's in Charge?"

26. Phil Plait, "The 50th Anniversary of Starfish Prime: The Nuke that Shook the
    World," *Discover*, July 9, 2012, https://www.discovermagazine.com/the-sciences
    /the-50th-anniversary-of-starfish-prime-the-nuke-that-shook-the-world.

27. Ian A. Crawford and Stephen Baxter, "The Lethality of Interplanetary Warfare:
    A Fundamental Constraint on Extraterrestrial Liberty," in *The Meaning of Liberty
    Beyond Earth*, ed. Charles S. Cockell (Cham, Switzerland: Springer, 2015), 192.

28. Stephen Baxter, "Architects of the Revolution: Design Philosophies to Maximise
    Extraterrestrial Liberty," in *Dissent, Revolution and Liberty Beyond Earth*, ed.
    Charles S. Cockell (Cham, Switzerland: Springer, 2016), 113.

29. Ian A. Crawford and Stephen Baxter, "The Lethality of Interplanetary Warfare,"
    119.

30. Brian Ferguson, "Ep 06: Who's in Charge?," interview by Erika Nesvold, *Making
    New Worlds*, December 20, 2017, podcast audio and transcript, https://making
    newworlds.com/2017/12/20/ep-06-whos-in-charge/.

31. Persson, "Citizens of Mars Ltd.," 122.

32. Charles S. Cockell, "Introduction: Human Governance and Liberty Beyond
    Earth," in *Human Governance Beyond Earth: Implications for Freedom*, ed. Charles
    S. Cockell (Cham, Switzerland: Springer, 2015), 4.

33. Baxter, "Architects of the Revolution," 111.

34. Paul Rosenberg, "What Happens to Dissent When Flight Is a Present Option?,"
    in *Dissent, Revolution and Liberty Beyond Earth*, ed. Charles S. Cockell (Cham,
    Switzerland: Springer, 2016), 155–167.

35. Charles S. Cockell, "Extraterrestrial Liberty: Can It Be Planned?," in *Human Gov-
    ernance Beyond Earth: Implications for Freedom*, ed. Charles S. Cockell (Cham,
    Switzerland: Springer, 2015), 39.

36. Benton, "Ep 06: Who's in Charge?"

37. Chris Newman, "Ep 06: Who's in Charge?," interview by Erika Nesvold, *Making New Worlds*, December 20, 2017, podcast audio and transcript, https://making newworlds.com/2017/12/20/ep-06-whos-in-charge/.

38. Ferguson, "Ep 06: Who's in Charge?"

39. UN General Assembly, "Treaty on Principles Governing the Activities of States in the Exploration and Use of Outer Space."

CHAPTER 10

1. Shawna Pandya, "Ep 07: What If I Want to Have Kids?," interview by Erika Nesvold, *Making New Worlds*, January 3, 2018, podcast audio and transcript, https://makingnewworlds.com/2018/01/03/ep-07-what-if-i-want-to-have-kids/.

2. Walter J. Sapp et al., "Effects of Spaceflight on the Spermatogonial Population of Rat Seminiferous Epithelium," *FASEB Journal* 4, no. 1 (1990): 101–104, https://doi.org/10.1096/fasebj.4.1.2295370; Sayaka Wakayama et al., "Detrimental Effects of Microgravity on Mouse Preimplantation Development In Vitro," *PLoS ONE* 4, no. 8 (2009): e6753, https://doi.org/10.1371/journal.pone.0006753; April E. Ronca et al., "Orbital Spaceflight During Pregnancy Shapes Mammalian Vestibular System," *Behavioral Neuroscience* 122, no. 1 (2008): 224–232, https://doi.org/10.1037/0735-7044.122.1.224.

3. Sayaka Wakayama et al., "Healthy Offspring from Freeze-Dried Mouse Spermatozoa Held on the International Space Station for 9 Months," *PNAS* 114, no. 23 (2017): 5988–5993, https://doi.org/10.1073/pnas.1701425114; Hans-Jürg Marthy, Philippe Schatt, and Luigia Santella, "Fertilization of Sea Urchin Eggs in Space and Subsequent Development under Normal Conditions," *Advances in Space Research* 14, no. 8 (1994): 197–208, https://doi.org/10.1016/0273-1177(94)90404-9; Kenichi Ijiri, "Fish Mating Experiment in Space—What It Aimed at and How It Was Prepared," *Biological Sciences in Space* 9, no. 1 (1995): 3–16, https://doi.org/10.2187/bss.9.3.

4. Nadia Drake, "Can Humans Have Babies on Mars? It May Be Harder Than You Think," *National Geographic*, December 10, 2018, https://www.nationalgeographic.com/science/2018/12/can-humans-have-babies-on-mars-space-it-may-be-harder-than-you-think/.

5. Dan Kabonge Kaye, "The Moral Imperative to Approve Pregnant Women's Participation in Randomized Clinical Trials for Pregnancy and Newborn Complications," *Philosophy, Ethics, and Humanities in Medicine* 14 (2019), https://doi.org/10.1186/s13010-019-0081-8.

6. Quill Kukla (as Rebecca Kukla), "Ep 07: What If I Want to Have Kids?," interview by Erika Nesvold, *Making New Worlds*, January 3, 2018, podcast audio and transcript, https://makingnewworlds.com/2018/01/03/ep-07-what-if-i-want-to-have-kids/.

7. Pandya, "Ep 07: What If I Want to Have Kids?"

8. Kukla, "Ep 07: What If I Want to Have Kids?"

9. David Perry, "The Ruderman White Paper: On Media Coverage of the Murder of People with Disabilities by Their Caregivers," Ruderman Family Foundation, March 2017, https://rudermanfoundation.org/white_papers/media-coverage-of-the-murder-of-people-with-disabilities-by-their-caregivers/.

10. A. H. Bittles and E. J. Glasson, "Clinical, Social, and Ethical Implications of Changing Life Expectancy in Down Syndrome," *Developmental Medicine & Child Neurology* 46, no. 4 (2004): 282–286, https://doi.org/10.1111/j.1469-8749.2004.tb00483.x.

11. Julian Quinones and Arijeta Lajka, "'What Kind of Society Do You Want to Live In?': Inside the Country Where Down Syndrome Is Disappearing," *CBS News*, August 14, 2017, https://www.cbsnews.com/news/down-syndrome-iceland/.

12. Jaime L. Natoli et al., "Prenatal Diagnosis of Down Syndrome: A Systematic Review of Termination Rates (1995–2011)," *Prenatal Diagnosis* 32, no. 2 (2012): 142–153, https://doi.org/10.1002/pd.2910.

13. Kukla, "Ep 07: What If I Want to Have Kids?"

14. Matthew Connelly, *Fatal Misconception: The Struggle to Control World Population* (Cambridge, MA: Belknap Press, 2008).

15. Connelly, *Fatal Misconception*.

16. Prajakta R. Gupte, "India: 'The Emergency' and the Politics of Mass Sterilization," *Education About Asia* 22, no. 3 (2017): 40–44, https://www.asianstudies.org/wp-content/uploads/india-the-emergency-and-the-politics-of-mass-sterilization.pdf.

17. Connelly, *Fatal Misconception*.

18. Kukla, "Ep 07: What If I Want to Have Kids?"

19. Connelly, *Fatal Misconception*.

20. Alexandra Minna Stern, "Forced Sterilization Policies in the US Targeted Minorities and Those with Disabilities—and Lasted into the 21st Century," *The Conversation*, August 26, 2020, https://theconversation.com/forced-sterilization-policies-in-the-us-targeted-minorities-and-those-with-disabilities-and-lasted-into-the-21st-century-143144.

21. Bibia Pavard, "The Right to Know? The Politics of Information about Contraception in France (1950s–'80s)," *Medical History* 63, no. 2 (2019): 173–188, https://dx.doi.org/10.1017%2Fmdh.2019.4.

22. Connelly, *Fatal Misconception*.

23. Kukla, "Ep 07: What If I Want to Have Kids?"

24. Cornelia Dean, "After a Struggle, Women Win a Place 'on the Ice'; In Labs and in the Field, a New Outlook," *New York Times*, November 10, 1998, https://www.nytimes.com/1998/11/10/science/after-struggle-women-win-place-ice-labs-field-new-outlook.html.

25. Ann Gibbons, "Sexual Harassment Is Common in Scientific Fieldwork," *Science*, July 16, 2014, https://www.sciencemag.org/news/2014/07/sexual-harassment-common-scientific-fieldwork.

26. Kukla, "Ep 07: What If I Want to Have Kids?"

27. Adrian LeBlanc et al., "Muscle Volume, MRI Relaxation Times (T2), and Body Composition After Spaceflight," *Journal of Applied Physiology* 89, no. 6 (2000): 2158–2164, https://doi.org/10.1152/jappl.2000.89.6.2158.

28. Pandya, "Ep 07: What If I Want to Have Kids?"

29. Connelly, *Fatal Misconception*, xi.

CHAPTER 11

1. Philippa Foot, "The Problem of Abortion and the Doctrine of the Double Effect," *Oxford Review* 5 (1967): 5–15, https://philarchive.org/archive/FOOTPO-2v1.

2. Kenneth V. Iserson, "Ep 10: What If I Get Sick?," interview by Erika Nesvold, *Making New Worlds*, January 31, 2018, podcast audio and transcript, https://makingnewworlds.com/2018/01/31/episode-10-what-if-i-get-sick/.

3. L. I. Rogozov, "Self Operation," *Soviet Antarctic Expedition Information Bulletin* 4 (1964): 223–224, English translation, https://www.southpolestation.com/trivia/igy1/rogozov.pdf.

4. Jim Duff, "Ep 10: What If I Get Sick?," interview by Erika Nesvold, *Making New Worlds*, January 31, 2018, podcast audio and transcript, https://makingnewworlds.com/2018/01/31/episode-10-what-if-i-get-sick/.

5. John Harris, *The Value of Life: An Introduction to Medical Ethics* (London: Routledge, 2006), 101.

6. American Thoracic Society Bioethics Task Force, "Fair Allocation of Intensive Care Unit Resources," *American Journal of Respiratory and Critical Care Medicine* 156, no. 4 (1997): 1282–1301, https://doi.org/10.1164/ajrccm.156.4.ats7-97.

7. Kenneth V. Iserson and John C. Moskop, "Triage in Medicine, Part I: Concept, History, and Types," *Annals of Emergency Medicine* 49, no. 3 (2007): 275–281, https://doi.org/10.1016/j.annemergmed.2006.05.019.

8. Keith Abney, "Ep 10: What If I Get Sick?," interview by Erika Nesvold, *Making New Worlds*, January 31, 2018, podcast audio and transcript, https://makingnewworlds.com/2018/01/31/episode-10-what-if-i-get-sick/.

9. International Committee of the Red Cross, *Convention (I) for the Amelioration of the Condition of the Wounded and Sick in Armed Forces in the Field*, August 12, 1949, 75 U.N.T.S. 31.

10. T. M. Egan et al., "Development of the New Lung Allocation System in the United States," *American Journal of Transplantation* 6, no. 5p2 (2006): 1212–1227, https://doi.org/10.1111/j.1600-6143.2006.01276.x.

11. Organ Procurement and Transplantation Network Ethics Committee, *Ethical Principles to Be Considered in the Allocation of Human Organs*, OPTN White Paper, 2010, https://optn.transplant.hrsa.gov/resources/ethics/ethical-principles-in-the-allocation-of-human-organs/.

12. Abney, "Ep 10: What If I Get Sick?"

13. Iserson, "Ep 10: What If I Get Sick?"

14. Organ Procurement and Transplantation Network Ethics Committee, *Ethical Principles to be Considered.*

15. Govind Persad, Alan Wertheimer, and Ezekiel J. Emanuel, "Principles for Allocation of Scarce Medical Interventions," *The Lancet* 373, no. 9661 (2009): 423–431, https://doi.org/10.1016/S0140-6736(09)60137-9.

16. US National Center for Health Workforce Analysis, "Sex, Race, and Ethnic Diversity of US Health Occupations (2011–2015)," Washington, DC: US Department of Health and Human Services, August 2017, https://bhw.hrsa.gov/sites/default/files/bureau-health-workforce/data-research/diversity-us-health-occupations.pdf.

17. Abney, "Ep 10: What If I Get Sick?"

18. Iserson, "Ep 10: What If I Get Sick?"

19. Robert M. Veatch, "Disaster Preparedness and Triage: Justice and the Common Good," *Mount Sinai Journal of Medicine* 72, no. 4 (2005): 236–241, https://www.researchgate.net/publication/7724070_Disaster_Preparedness_and_Triage_Justice_and_the_Common_Good.

20. Elizabeth L. Daugherty Biddison et al., "The Community Speaks: Understanding Ethical Values in Allocation of Scarce Livesaving Resources during Disasters," *Annals of the American Thoracic Society* 11, no. 5 (2014), 777–783, https://doi.org/10.1513/AnnalsATS.201310-379OC.

21. Abney, "Ep 10: What If I Get Sick?"

CHAPTER 12

1. Frank White, *The Overview Effect: Space Exploration and Human Evolution* (Boston: Houghton-Mifflin, 1987).

2. Alice Qannik Glenn, "Ep 11: Which Way Is Mecca?," interview by Erika Nesvold, *Making New Worlds*, February 7, 2018, podcast audio and transcript, https://makingnewworlds.com/2018/02/07/ep-11-which-way-is-mecca/.

3. Michael Waltemathe, "Ep 11: Which Way Is Mecca?," interview by Erika Nesvold, *Making New Worlds*, February 7, 2018, podcast audio and transcript, https://makingnewworlds.com/2018/02/07/ep-11-which-way-is-mecca/.

4. Deana Weibel, interview with the author, April 2, 2020.

5. Waltemathe, "Ep 11: Which Way Is Mecca?"

6. Waltemathe, "Ep 11: Which Way Is Mecca?"

7. Mehmet Ozalp, "Ep 11: Which Way Is Mecca?," interview by Erika Nesvold, *Making New Worlds*, February 7, 2018, podcast audio and transcript, https://makingnewworlds.com/2018/02/07/ep-11-which-way-is-mecca/.

8. Department of Islamic Development Malaysia, "Guideline for Performing Islamic Rites at the International Space Station (ISS)," 2007, https://www.islam
.gov.my/images/garis-panduan/Buku-Garis-Panduan-ISS-Dalam-Versi-Malaysia
-Arab-Russia-Inggeris.pdf.

9. Deana L. Weibel, "The Overview Effect and the Ultraview Effect: How Extreme Experiences in/of Outer Space Influence Religious Beliefs in Astronauts," *Religions* 11, no. 8 (2020): 418–434, https://www.doi.org/10.3390/rel11080418.

10. Weibel, interview.

11. Glenn, "Ep 11: Which Way Is Mecca?"

12. Ozalp, "Ep 11: Which Way Is Mecca?"

CHAPTER 13

1. James S. J. Schwartz and Tony Milligan, "'Space Ethics' According to Space Ethicists," *Space Review*, February 1, 2021, https://www.thespacereview.com/article
/4117/1; Hannah Arendt, *The Human Condition* (Chicago: University of Chicago Press, 1958), 1.

2. David Valentine, Valerie A. Olson, and Debbora Battaglia, "Encountering the Future: Anthropology and Outer Space," *Anthropology News* 50, no. 9 (2009): 11–15, https://doi.org/10.1111/j.1556-3502.2009.50911.x.

3. Kathryn Denning, "Is Life What We Make of It?," *Philosophical Transactions of the Royal Society A* 369, no. 1936 (2011): 669–678, https://doi.org/10.1098/rsta.2010.0230; Michael P. Oman-Reagan, "Queering Outer Space" (unpublished manuscript, July 2, 2018), SocArXiv, Open Science Framework, https://doi.org/10.31235/osf.io
/mpyk6.

4. Lisa Ruth Rand, "Falling Cosmos: Nuclear Reentry and the Environmental History of Earth Orbit," *Environmental History* 24, no. 1 (2019): 78–103, https://doi.org
/10.1093/envhis/emy125; Asif A. Siddiqi, "Competing Technologies, National(ist) Narratives, and Universal Claims: Toward a Global History of Space Exploration," *Technology and Culture* 51, no. 2 (2010): 425–443, https://doi.org/10.1353
/tech.0.0459.

5. Daniel Deudney, *Dark Skies: Space Expansionism, Planetary Geopolitics, and the Ends of Humanity* (New York: Oxford University Press, 2020); Fred Scharmen, *Space Forces: A Critical History of Life in Outer Space* (London: Verso, 2021).

6. Kyle Whyte, interview with the author, March 29, 2021.

BIBLIOGRAPHY

Abney, Keith. "Ep 10: What If I Get Sick?" Interview by Erika Nesvold. *Making New Worlds*, January 31, 2018. Podcast audio and transcript. https://makingnewworlds .com/2018/01/31/episode-10-what-if-i-get-sick/.

Adams, Douglas. *The Hitchhiker's Guide to the Galaxy*. New York: Harmony Books, 1980.

Alexander, Danielle. "Forty Acres and a Mule: The Ruined Hope of Reconstruction." *Humanities* 25, no. 1 (2004). https://web.archive.org/web/20080916095443/http://neh.gov /news/humanities/2004-01/reconstruction.html.

Alexander, Michelle. *The New Jim Crow: Mass Incarceration in the Age of Colorblindness*. New York: New Press, 2010.

American Thoracic Society Bioethics Task Force. "Fair Allocation of Intensive Care Unit Resources." *American Journal of Respiratory and Critical Care Medicine* 156, no. 4 (1997): 1282–1301. https://doi.org/10.1164/ajrccm.156.4.ats7-97.

Anderson, Ross. "Exodus: Elon Musk Argues that We Must Put a Million People on Mars If We Are to Ensure that Humanity Has a Future." *Aeon*, September 30, 2014. https://aeon.co/essays/elon-musk-puts-his-case-for-a-multi-planet-civilisation.

Arendt, Hannah. *The Human Condition*. Chicago: University of Chicago Press, 1958.

Arizona Correctional Industries. "About ACI." June 27, 2021. Wayback Machine. https:// web.archive.org/web/20210627235219/https://aci.az.gov/about/.

Arizona Correctional Industries. "About ACI." February 17, 2022. Wayback Machine. https://web.archive.org/web/20220217021127/https://aci.az.gov/about/.

Arneil, Barbara. "The Wild Indian's Venison: Locke's Theory of Property and English Colonialism in America." *Political Studies* 44, no. 1 (1996): 60–74. https://journals .sagepub.com/doi/10.1111/j.1467-9248.1996.tb00764.x.

Arnould, Jacques. *Icarus' Second Chance: The Basis and Perspectives of Space Ethics*. New York: SpringerWienNewYork, 2011.

Baker, David. "Liberty, Freedom and Democracy: Paradox for an Extraterrestrial Society." In *The Meaning of Liberty Beyond Earth*, edited by Charles S. Cockell, 227–249. Cham, Switzerland: Springer, 2015.

Bartels, Meghan. "Apollo 11 Astronauts Spent 3 Weeks in Quarantine, Just in Case of Moon Plague." *Space.com*, July 24, 2019. https://www.space.com/apollo-11-astronauts -quarantined-after-splashdown.html.

Barton, Susanne, and Hannah Recht. "The Massive Prize Luring Miners to the Stars." *Bloomberg*, March 8, 2018. https://www.bloomberg.com/graphics/2018-asteroid-mining/.

Baxter, Stephen. "Architects of the Revolution: Design Philosophies to Maximise Extraterrestrial Liberty." In *Dissent, Revolution and Liberty Beyond Earth*, edited by Charles S. Cockell, 111–124. Cham, Switzerland: Springer, 2016.

Baxter, Stephen. "The Cold Equations: Extraterrestrial Liberty in Science Fiction." In *The Meaning of Liberty Beyond Earth*, edited by Charles S. Cockell, 13–31. Cham, Switzerland: Springer, 2015.

Becher, Debbie. "Ep 03: Who Owns Mars?" Interview by Erika Nesvold. *Making New Worlds*, November 29, 2017. Podcast audio and transcript. https://makingnewworlds.com/2017/11/29/ep-03-who-owns-mars/.

Benton, Lauren. "Ep 06: Who's in Charge?" Interview by Erika Nesvold. *Making New Worlds*, December 20, 2017. Podcast audio and transcript. https://makingnewworlds.com/2017/12/20/ep-06-whos-in-charge/.

Bergreen, Laurence. *Over the Edge of the World: Magellan's Terrifying Circumnavigation of the Globe*. New York: Harper Collins, 2003.

Berry, Wendell. "Comments on O'Neill's Space Colonies." In *Space Colonies*, edited by Stewart Brand. San Francisco: Waller Press, 1977. https://space.nss.org/settlement/nasa/CoEvolutionBook/DEBATE.HTML.

Bhatt, Mukesh Chiman. "Space for Dissent: Disobedience on Artificial Habitats and Planetary Settlements." In *Dissent, Revolution and Liberty Beyond Earth*, edited by Charles S. Cockell, 71–92. Cham, Switzerland: Springer, 2016.

Billings, Linda. "Should Humans Colonize Other Planets? No." *Theology and Science* 15, no. 3 (2017): 321–332. https://doi.org/10.1080/14746700.2017.1335065.

Billings, Linda. "Space Cowboys: How Jingoism Corrupts American Rhetoric on Human Spaceflight." *Scientific American* 313, no. 2 (2015): 12. http://doi.org/10.1038/scientificamerican0815-12.

Bittles, A. H., and E. J. Glasson. "Clinical, Social, and Ethical Implications of Changing Life Expectancy in Down Syndrome." *Developmental Medicine & Child Neurology* 46, no. 4 (2004): 282–286. https://doi.org/10.1111/j.1469-8749.2004.tb00483.x.

Blackmon, Douglas A. *Slavery by Another Name: The Re-Enslavement of Black Americans from the Civil War to World War II*. New York: Doubleday, 2008.

Bloom, Ester. "Elon Musk's SpaceX Shortchanged Its Workers and Now It Must Pay." *CNBC*, May 15, 2017. https://www.cnbc.com/2017/05/15/elon-musks-spacex-mistreated-its-workers-and-now-it-must-pay.html.

Bobroff, Kenneth H. "Retelling Allotment: Indian Property Rights and the Myth of Common Ownership." *Vanderbilt Law Review* 54, no. 4 (2001): 1559–1623. https://scholarship.law.vanderbilt.edu/cgi/viewcontent.cgi?article=1879&context=vlr.

Bolden, Charles. "Sending Humans to Mars." Filmed April 22, 2014 in Washington, DC. Humans 2 Mars Summit video, 29:12. https://www.c-span.org/video/?318982-1 /human-mars-summit.

Brown, Michelle. "Ep 05: What If Someone Steals My Stuff?" Interview by Erika Nesvold. *Making New Worlds*, December 13, 2017. Podcast audio and transcript. https:// makingnewworlds.com/2017/12/13/ep-05-what-if-someone-steals-my-stuff/.

Brown, Patricia Leigh. "Prison Makes Way for Future, but Preserves Past." *San Quentin Journal*, January 18, 2008. https://www.nytimes.com/2008/01/18/us/18dungeon.html.

Brown Weiss, Edith. "In Fairness to Future Generations." *Environment: Science and Policy for Sustainable Development* 32, no. 3 (1990): 6–31. https://doi.org/10.1080/00139157 .1990.9929015.

Brown Weiss, Edith. "In Fairness to Future Generations and Sustainable Development." *American University International Law Review* 8, no. 1 (1992): 19–26. https:// digitalcommons.wcl.american.edu/auilr/vol8/iss1/2/.

Bucher, Enrique H. "The Causes of Extinction of the Passenger Pigeon." In *Current Ornithology*, vol. 9, edited by Dennis M. Power, 1–36. Boston: Springer, 1992. https:// link.springer.com/chapter/10.1007%2F978-1-4757-9921-7_1.

Bury, J. B. *The Idea of Progress*. London: Macmillan, 1920.

Cantor, Matthew. "NASA Cancels All-Female Spacewalk, Citing Lack of Spacesuit in Right Size." *The Guardian*, March 26, 2019. https://www.theguardian.com/science/2019 /mar/25/nasa-all-female-spacewalk-canceled-women-spacesuits.

Capper, Daniel. "Preserving Mars Today Using Baseline Ecologies." *Space Policy* 49 (2019). https://doi.org/10.1016/j.spacepol.2019.05.003.

Chang, Gordon H., Shelley Fisher Fishkin, and Hilton Obenzinger. Introduction to *The Chinese and the Iron Road: Building the Transcontinental Railroad*, 1–26. Edited by Gordon H. Chang and Shelley Fisher Fishkin. Stanford: Stanford University Press, 2019.

Chang, Kenneth. "A Billionaire Names His Team to Ride SpaceX, No Pros Allowed." *New York Times*, March 30, 2021. https://www.nytimes.com/2021/03/30/science/30spacex -inspiration4.html.

Chen, Chuansheng, Michael Burton, Ellen Greenberger, and Julia Dmitrieva. "Population Migration and the Variation of Dopamine D4 Receptor (DRD4) Allele Frequencies Around the Globe." *Evolution and Human Behavior* 20, no. 5 (1999): 309–324. https:// doi.org/10.1016/S1090-5138(99)00015-X.

Cockell, Charles S. "Disobedience in Outer Space." In *Dissent, Revolution and Liberty Beyond Earth*, edited by Charles S. Cockell, 21–40. Cham, Switzerland: Springer, 2016.

Cockell, Charles S. "Exoconfac—the Extraterrestrial Containment Facility: An Essay on the Design Philosophy for a Prison to Contain Criminals in Settlements Beyond the Earth." *Journal of the British Interplanetary Society* 69 (2016): 27–30. https://www

.researchgate.net/publication/306208635_Exoconfac_-_The_extraterrestrial_contain ment_facility_An_essay_on_the_design_philosophy_for_a_prison_to_contain _criminals_in_settlements_beyond_the_earth.

Cockell, Charles S. "Extraterrestrial Liberty: Can It Be Planned?" In *Human Governance Beyond Earth: Implications for Freedom*, edited by Charles S. Cockell, 23–42. Cham, Switzerland: Springer, 2015.

Cockell, Charles S. "Freedom in a Box: Paradoxes in the Structure of Extraterrestrial Liberty." In *The Meaning of Liberty Beyond Earth*, edited by Charles S. Cockell, 47–68. Cham, Switzerland: Springer, 2015.

Cockell, Charles S. "Introduction: Human Governance and Liberty Beyond Earth." In *Human Governance Beyond Earth: Implications for Freedom*, edited by Charles S. Cockell, 1–8. Cham, Switzerland: Springer, 2015.

Cockell, Charles S., and Gerda Horneck. "A Planetary Park System for Mars." *Space Policy* 20, no. 4 (2004): 291–295. https://doi.org/10.1016/j.spacepol.2004.08.003.

Cockell, Charles S., and Gerda Horneck. "Planetary Parks—Formulating a Wilderness Policy for Planetary Bodies." *Space Policy* 22, no. 4 (2006): 256–261. https://doi.org /10.1016/j.spacepol.2006.08.006.

Connelly, Matthew. *Fatal Misconception: The Struggle to Control World Population*. Cambridge, MA: Belknap Press, 2008.

Cooke, Brian D. "Rabbits: Manageable Environmental Pests or Participants in New Australian Ecosystems?" *Wildlife Research* 39, no. 4 (2012): 279–289. http://doi.org/10.1071 /WR11166.

COSPAR Panel on Planetary Protection. "COSPAR Policy on Planetary Protection." June 17, 2020. https://cosparhq.cnes.fr/assets/uploads/2020/07/PPPolicyJune-2020 _Final_Web.pdf.

Crawford, Ian A., and Stephen Baxter. "The Lethality of Interplanetary Warfare: A Fundamental Constraint on Extraterrestrial Liberty." In *The Meaning of Liberty Beyond Earth*, edited by Charles S. Cockell, 187–198. Cham, Switzerland: Springer, 2015.

Daugherty Biddison, Elizabeth L., Howard Gwon, Monica Schoch-Spana, Robert Cavalier, Douglas B. White, Timothy Dawson, Peter B. Terry, Alex John London, Alan Regenberg, Ruth Faden, and Eric S. Toner. "The Community Speaks: Understanding Ethical Values in Allocation of Scarce Livesaving Resources during Disasters." *Annals of the American Thoracic Society* 11, no. 5 (2014): 777–783. https://doi.org/10.1513 /AnnalsATS.201310-379OC.

Davenport, Christian. "Elon Musk on Mariachi Bands, Zero-G Games, and Why His Mars Plan Is Like 'Battlestar Galactica.'" *Washington Post*, September 28, 2016. https:// www.washingtonpost.com/news/the-switch/wp/2016/09/28/elon-musk-on-mariachi -bands-how-he-plans-to-make-space-travel-fun-and-why-his-mars-plan-is-like-battlestar -galactica/.

Dean, Cornelia. "After a Struggle, Women Win a Place 'on the Ice'; In Labs and in the Field, a New Outlook." *New York Times*, November 10, 1998. https://www.nytimes .com/1998/11/10/science/after-struggle-women-win-place-ice-labs-field-new-outlook .html.

Declaration of the First Meeting of Equatorial Countries. Bra.-Col.-Cog.-Ecu.-Idn. -Ken.-Uga.-Zar. December 3, 1976. https://www.jaxa.jp/library/space_law/chapter_2/2-2 -1-2_e.html.

Denevan, William M. "The Pristine Myth: The Landscape of the Americas in 1492." *Annals of the Association of American Geographers* 82, no. 3 (1992): 369–385. https://doi .org/10.1111/j.1467-8306.1992.tb01965.x.

Denning, Kathryn. "Is Life What We Make of It?" *Philosophical Transactions of the Royal Society A* 369, no. 1936 (2011): 669–678. https://doi.org/10.1098/rsta.2010.0230.

Department of Islamic Development Malaysia. "Guideline for Performing Islamic Rites at the International Space Station (ISS)." 2007. https://www.islam.gov.my/images /garis-panduan/Buku-Garis-Panduan-ISS-Dalam-Versi-Malaysia-Arab-Russia-Inggeris.pdf.

Deudney, Daniel. *Dark Skies: Space Expansionism, Planetary Geopolitics, and the Ends of Humanity*. New York: Oxford University Press, 2020.

Dieker, Nicole. "Mars: The Rich Planet." *Medium*, September 28, 2016. https://medium .com/the-billfold/mars-the-rich-planet-e6be9836of2b.

Dixon, Bernard. "Smallpox—Imminent Extinction, and an Unresolved Dilemma." *New Scientist* 69, no. 989 (1976): 430–432.

Dourado, Eli (Senior Research Fellow and Director, Technology Policy Program, Mercatus Center, George Mason University). *Regulating Space: Innovation, Liberty, and International Obligations: Hearing before the Science, Space, and Technology Subcommittee on Space, House of Representatives*. 115th Cong. (March 8, 2017). https://www.mercatus .org/publications/technology-and-innovation/creating-environment-permissionless -innovation-outer-space.

Drake, Nadia. "Can Humans Have Babies on Mars? It May Be Harder Than You Think." *National Geographic*, December 10, 2018. https://www.nationalgeographic.com/science /2018/12/can-humans-have-babies-on-mars-space-it-may-be-harder-than-you-think/.

Duff, Jim. "Ep 10: What If I Get Sick?" Interview by Erika Nesvold. *Making New Worlds*, January 31, 2018. Podcast audio and transcript. https://makingnewworlds .com/2018/01/31/episode-10-what-if-i-get-sick/.

Egan, T. M., S. Murray, R. T. Bustami, T. H. Shearon, K. P. McCullough, L. B. Edwards, M. A. Coke, E. R. Garrity, S. C. Sweet, D. A. Heiney, and F. L. Grover. "Development of the New Lung Allocation System in the United States." *American Journal of Transplantation* 6, no. 5p2 (2006): 1212–1227. https://doi.org/10.1111/j.1600-6143.2006.01276.x.

Eisen, Lauren-Brooke. "Covid-19 Highlights the Need for Prison Labor Reform." *Brennan Center for Justice*, April 17, 2020. https://www.brennancenter.org/our-work /analysis-opinion/covid-19-highlights-need-prison-labor-reform.

Eisner, Breck, dir. *The Expanse*. Season 4, episode 10, "Cibola Burn." Aired December 13, 2019, on Amazon Prime.

Elvis, Martin. "How Many Ore-Bearing Asteroids?" *Planetary and Space Science* 91 (2014): 20–26. http://doi.org/10.1016/j.pss.2013.11.008.

Elvis, Martin, Alanna Krolikowski, and Tony Milligan. "Concentrated Lunar Resources: Imminent Implications for Governance and Justice." *Philosophical Transactions of the Royal Society A* 379, no. 2188 (2020). http://doi.org/10.1098/rsta.2019.0563.

Elvis, Martin, and Tony Milligan. "How Much of the Solar System Should We Leave as Wilderness?" *Acta Astronautica* 162 (2019): 574–580. https://doi.org/10.48550/arXiv .1905.13681.

Elvis, Martin, Tony Milligan, and Alanna Krolikowski. "The Peaks of Eternal Light: A Near-Term Property Issue on the Moon." *Space Policy* 38 (2016): 30–38. http://doi .org/10.1016/j.spacepol.2016.05.011.

Ervin, Scott. "Law in a Vacuum: The Common Heritage Doctrine in Outer Space Law." *Boston College International and Comparative Law Review* 7, no. 2 (1984): 403–431. http:// lawdigitalcommons.bc.edu/iclr/vol7/iss2/9.

European Space Agency. "ESA Commissions World's First Space Debris Removal." Accessed September 12, 2019. https://www.esa.int/Safety_Security/Clean_Space/ESA _commissions_world_s_first_space_debris_removal.

European Space Agency. "Parastronaut Feasibility Project." Accessed April 3, 2021. https://www.esa.int/About_Us/Careers_at_ESA/ESA_Astronaut_Selection/Paras tronaut_feasibility_project.

European Space Agency. "Psychological and Medical Selection Process." Accessed April 3, 2021. https://www.esa.int/Science_Exploration/Human_and_Robotic_Explora tion/European_Astronaut_Selection_2008/Psychological_and_medical_selection _process.

European Space Agency. "Space Debris by the Numbers." Accessed November 1, 2020. https://www.esa.int/Safety_Security/Space_Debris/Space_debris_by_the_numbers.

Ferguson, Brian. "Ep 06: Who's in Charge?" Interview by Erika Nesvold. *Making New Worlds*, December 20, 2017. Podcast audio and transcript. https://makingnewworlds .com/2017/12/20/ep-06-whos-in-charge/.

Fernholz, Tim. "The US Government Has Approved the First Private Landing on the Moon." *Quartz*, August 3, 2016. https://qz.com/749246/the-us-government-has -approved-the-first-private-landing-on-the-moon/.

Fogg, Martyn J. "The Ethical Dimensions of Space Settlement." *Space Policy* 16, no. 3 (2000): 205–211. http://doi.org/10.1016/S0265-9646(00)00024-2.

Foot, Philippa. "The Problem of Abortion and the Doctrine of the Double Effect." *Oxford Review* 5 (1967): 5–15. https://philarchive.org/archive/FOOTPO-2v1.

Foust, Jeff. "Bigelow to Press US Government on Lunar Property Rights." *Space Politics*, November 13, 2013. http://www.spacepolitics.com/2013/11/13/bigelow-to-press-us-government-on-lunar-property-rights/.

Foust, Jeff. "The Cosmic Vision of Jeff Bezos." *SpaceNews*, February 25, 2019. https://spacenews.com/the-cosmic-vision-of-jeff-bezos/.

Foust, Jeff. "New Law Unlikely to Settle Debate on Space Resource Rights." *SpaceNews*, December 4, 2015. https://spacenews.com/new-law-unlikely-to-settle-debate-on-space-resource-rights/.

Frankema, E. H. P. "The Colonial Origins of Inequality: Exploring the Causes and Consequences of Land Distribution." *IAI Discussion Papers*, no. 119 (2005). https://www.econstor.eu/bitstream/10419/27410/1/504473565.PDF.

French, Robert Heath. "Environmental Philosophy and the Ethics of Terraforming Mars: Adding the Voices of Environmental Justice and Ecofeminism to the Ongoing Debate." Master's thesis, University of North Texas, August 2013.

Friedrich, Otto. *The Kingdom of Auschwitz*. New York: HarperCollins, 1994.

Gabaccia, Donna. "Ep 01: Why Are We Going?" Interview by Erika Nesvold. *Making New Worlds*, November 15, 2017. Podcast audio and transcript. https://makingnewworlds.com/2017/11/15/episode-1-why-are-we-going/.

Gibbons, Ann. "Sexual Harassment Is Common in Scientific Fieldwork." *Science*, July 16, 2014. https://www.sciencemag.org/news/2014/07/sexual-harassment-common-scientific-fieldwork.

Glenn, Alice Qannik. "Ep 11: Which Way Is Mecca?" Interview by Erika Nesvold. *Making New Worlds*, February 7, 2018. Podcast audio and transcript. https://makingnewworlds.com/2018/02/07/ep-11-which-way-is-mecca/.

Global Justice Now. "69 of the Richest 100 Entities on the Planet Are Corporations, Not Governments, Figures Show." October 17, 2018. https://www.globaljustice.org.uk/news/2018/oct/17/69-richest-100-entities-planet-are-corporations-not-governments-figures-show.

Godwin, Tom. "The Cold Equations." *Astounding Science Fiction* (August 1954). Reprinted in *Lightspeed Magazine* 14 (July 2011). https://www.lightspeedmagazine.com/fiction/the-cold-equations/.

Gonzales-Day, Ken. *Lynching in the West: 1850–1935*. Durham, NC: Duke University Press, 2006.

Gore, Al. *Earth in the Balance: Ecology and the Human Spirit*. New York: Houghton Mifflin, 1992.

Gorman, Alice. "Can the Moon Be a Person? As Lunar Mining Looms, a Change of Perspective Could Protect Earth's Ancient Companion." *The Conversation*, August 26, 2020. https://theconversation.com/can-the-moon-be-a-person-as-lunar-mining-looms-a-change-of-perspective-could-protect-earths-ancient-companion-144848.

Green, J. L., J. Hollingsworth, D. Brain, V. Airapetian, A. Glocer, A. Pulkkinen, C. Dong, and R. Bamford. "A Future Mars Environment for Science and Exploration." Paper presented at the Planetary Science Vision 2050 Workshop, Washington, DC, February 2017. https://www.hou.usra.edu/meetings/V2050/pdf/8250.pdf.

Griffin, Michael D. "Chapter 1." In *NASA at 50: Interviews with NASA's Senior Leadership*, edited by Rebecca Wright, Sandra Johnson, and Steven J. Dick, 1–23. Washington, DC: US Government Printing Office, 2012.

Grubb, Farley. "The Incidence of Servitude in Trans-Atlantic Migration, 1771–1804." *Explorations in Economic History* 22, no. 3 (1985): 316–339. https://doi.org/10.1016/0014 -4983(85)90016-6.

Grush, Loren. "Jeff Bezos' Space Company Is Pressuring Employees to Launch a Tourist Rocket during the Pandemic." *The Verge*, April 2, 2020. https://www.theverge .com/2020/4/2/21198272/blue-origin-coronavirus-leaked-audio-test-launch-workers-jeff -bezos.

Grush, Loren. "NASA Announces International Artemis Accords to Standardize How to Explore the Moon." *The Verge*, May 15, 2020. https://www.theverge.com/2020/5/15 /21259946/nasa-artemis-accords-lunar-exploration-moon-outer-space-treaty.

Gupte, Prajakta R. "India: 'The Emergency' and the Politics of Mass Sterilization." *Education About Asia* 22, no. 3 (2017): 40–44. https://www.asianstudies.org/wp-content /uploads/india-the-emergency-and-the-politics-of-mass-sterilization.pdf.

Hardin, Garrett. "Lifeboat Ethics: The Case Against Helping the Poor." *Psychology Today Magazine* 8 (1974): 38–43. http://www.garretthardinsociety.org/articles/art_life boat_ethics_case_against_helping_poor.html.

Hardin, Garrett. "Living on a Lifeboat." *BioScience* 24, no. 10 (1974): 561–568. http:// www.garretthardinsociety.org/articles/art_living_on_a_lifeboat.html.

Hardin, Garrett. "The Tragedy of the Commons." *Science* 162, no. 3859 (1968): 1243–1248. http://doi.org/10.1126/science.162.3859.1243.

Harrison, Albert A. "Russian and American Cosmism: Religion, National Psyche, and Spaceflight." *Astropolitics* 11, no. 1–2 (2013): 25–44. http://doi.org/10.1080/14777622.2013 .801719.

Hartmann, William K. "Space Exploration and Environmental Issues." In *Beyond Spaceship Earth: Environmental Ethics and the Solar System*, edited by Eugene C. Hargrove, 119–139. San Francisco: Sierra Club Books, 1986.

Haskins, Caroline. "An Attempted Murder at a Research Station Shows How Crimes Are Prosecuted in Antarctica." *Vice*, October 25, 2018. https://www.vice.com/en_us /article/xw9bg3/an-attempted-murder-at-a-research-station-shows-how-crimes-are -prosecuted-in-antarctica.

Herbert, Nick. "The Abolitionists' Criminal Conspiracy." *The Guardian*, July 27, 2008. https://www.theguardian.com/commentisfree/2008/jul/27/prisonsandprobation .youthjustice.

Hertzfeld, Henry R., Brian Weeden, and Christopher D. Johnson. "How Simple Terms Mislead Us: The Pitfalls of Thinking about Outer Space as a Commons." *International Astronautical Congress 15*, 2015. https://swfound.org/media/205390/how-simple-terms -mislead-us-hertzfeld-johnson-weeden-iac-2015.pdf.

Hiltzik, Michael. "The Day When Three NASA Astronauts Staged a Strike in Space." *Los Angeles Times*, December 28, 2015. https://www.latimes.com/business/hiltzik/la-fi -mh-that-day-three-nasa-astronauts-20151228-column.html.

Hiscox, Julian A., and David J. Thomas. "Genetic Modification and Selection of Micro-organisms for Growth on Mars." *Journal of the British Interplanetary Society* 48, no. 10 (1995): 419–426. https://pubmed.ncbi.nlm.nih.gov/11541203/.

Holthaus, Eric. *The Future Earth: A Radical Vision for What's Possible in the Age of Warming*. New York: HarperCollins, 2020.

Hui, Sylvia. "Hawking: Humans Must Spread Out in Space." *Associated Press*, June 13, 2006. https://www.washingtonpost.com/wp-dyn/content/article/2006/06/13/AR200606 1301185_pf.html.

Hume, Ivor Noël. "We Are Starved." *Colonial Williamsburg Journal* 29, no. 1 (2007): 44–51. https://research.colonialwilliamsburg.org/Foundation/journal/Winter07/starving .cfm.

Ijiri, Kenichi. "Fish Mating Experiment in Space—What It Aimed at and How It Was Prepared." *Biological Sciences in Space* 9, no. 1 (1995): 3–16. https://doi.org/10.2187 /bss.9.3.

Imarisha, Walidah. "Ep 05: What If Someone Steals My Stuff?" Interview by Erika Nesvold. *Making New Worlds*, December 13, 2017. Podcast audio and transcript. https:// makingnewworlds.com/2017/12/13/ep-05-what-if-someone-steals-my-stuff/.

International Committee of the Red Cross. *Convention (I) for the Amelioration of the Condition of the Wounded and Sick in Armed Forces in the Field*. August 12, 1949, 75 U.N.T.S. 31.

Iserson, Kenneth V. "Ep 10: What If I Get Sick?" Interview by Erika Nesvold. *Making New Worlds*, January 31, 2018. Podcast audio and transcript. https://makingnewworlds .com/2018/01/31/episode-10-what-if-i-get-sick/.

Iserson, Kenneth V., and John C. Moskop. "Triage in Medicine, Part I: Concept, History, and Types." *Annals of Emergency Medicine* 49, no. 3 (2007): 275–281. https:// doi.org/10.1016/j.annemergmed.2006.05.019.

Islam, S. Nazrul, and John Winkel. "Climate Change and Social Inequality." DESA Working Paper No. 152 ST/ESA/2017/DWP/152. United Nations Department of Economic & Social Affairs, 2017. https://www.un.org/esa/desa/papers/2017/wp152_2017.pdf.

Kaye, Dan Kabonge. "The Moral Imperative to Approve Pregnant Women's Participation in Randomized Clinical Trials for Pregnancy and Newborn Complications." *Philosophy, Ethics, and Humanities in Medicine* 14 (2019). https://doi.org/10.1186 /s13010-019-0081-8.

Kennedy, John F. "Address at Rice University on the Nation's Space Effort." Filmed September 12, 1962, in Houston. Historic speech transcript and video, 18:27. https://www.jfklibrary.org/learn/about-jfk/historic-speeches/address-at-rice-university-on-the-nations-space-effort.

Kessler, Donald J., and Burton G. Cour-Palais. "Collision Frequency of Artificial Satellites: The Creation of a Debris Belt." *Journal of Geophysical Research* 83, no. A6 (1978): 2637–2646. https://doi.org/10.1029/JA083iA06p02637.

Kramer, Katie. "Neil deGrasse Tyson Says Space Ventures Will Spawn First Trillionaire." *NBC News*, May 3, 2015. https://www.nbcnews.com/science/space/neil-degrasse-tyson-says-space-ventures-will-spawn-first-trillionaire-n352271.

Krznaric, Roman. *The Good Ancestor: A Radical Prescription for Long-Term Thinking.* New York: The Experiment, 2020.

Kukla, Quill (as Rebecca Kukla). "Ep 07: What If I Want to Have Kids?" Interview by Erika Nesvold. *Making New Worlds*, January 3, 2018. Podcast audio and transcript. https://makingnewworlds.com/2018/01/03/ep-07-what-if-i-want-to-have-kids/.

Launius, Roger D. "Escaping Earth: Human Spaceflight as Religion." *Astropolitics* 11, no. 1–2 (2013): 45–64. http://doi.org/10.1080/14777622.2013.801720.

Launius, Roger D. "The Railroads and the Space Program Revisited: Historical Analogues and the Stimulation of Commercial Space Operations." *Astropolitics* 12, no. 2–3 (2014): 167–179. https://doi.org/10.1080/14777622.2014.964129.

LeBlanc, Adrian, Chen Lin, Linda Shackelford, Valentine Sinitsyn, Harlan Evans, Oleg Belichenko, Boris Schenkman, Inessa Kozlovskaya, Victor Oganov, Alexi Bakulin, Thomas Hedrick, and Daniel Feeback. "Muscle Volume, MRI Relaxation Times (T2), and Body Composition after Spaceflight." *Journal of Applied Physiology* 89, no. 6 (2000): 2158–2164. https://doi.org/10.1152/jappl.2000.89.6.2158.

Leopold, Aldo. "The Land Ethic." In *A Sand County Almanac: And Sketches Here and There*, 201–226. Oxford: Oxford University Press, 1949.

Letizia, Francesca, Stijn Lemmens, Benjamin Bastida Virgili, and Holger Krag. "Application of a Debris Index for Global Evaluation of Mitigation Strategies." *Acta Astronautica* 161 (2019): 348–362. https://doi.org/10.1016/j.actaastro.2019.05.003.

Little Badger, Darcie. "Ep 01: Why Are We Going?" Interview by Erika Nesvold. *Making New Worlds*, November 15, 2017. Podcast audio and transcript. https://makingnewworlds.com/2017/11/15/episode-1-why-are-we-going/.

Lowe, Jaime. "The Incarcerated Women Who Fight California's Wildfires." *New York Times Magazine*, August 31, 2017. https://www.nytimes.com/2017/08/31/magazine/the-incarcerated-women-who-fight-californias-wildfires.html.

Lupisella, Mark, and John Logsdon. "Do We Need a Cosmocentric Ethic?" Paper IAA-97-IAA.9.2.09 presented at the International Astronautical Federation Congress, Turin, Italy, 1997. https://www.academia.edu/266597/Do_We_Need_a_Cosmocentric_Ethic.

Lurgio, Jeremy. "Saving the Whanganui: Can Personhood Rescue a River?" *The Guardian*, November 29, 2019. https://www.theguardian.com/world/2019/nov/30/saving-the-whanganui-can-personhood-rescue-a-river.

Lyons, Oren. "An Iroquois Perspective." In *American Indian Environments: Ecological Issues in Native American History*, edited by Christopher Vecsey and Robert W. Venables, 171–175. Syracuse, NY: Syracuse University Press, 1980.

Mann, Charles C. *1491: New Revelations of the Americas Before Columbus*. 2nd ed. New York: Vintage, 2006.

Marshall, Logan, ed. *On Board the Titanic: The Complete Story with Eyewitness Accounts*. Mineola, NY: Dover, 2006.

Marthy, Hans-Jürg, Philippe Schatt, and Luigia Santella. "Fertilization of Sea Urchin Eggs in Space and Subsequent Development under Normal Conditions." *Advances in Space Research* 14, no. 8 (1994): 197–208. https://doi.org/10.1016/0273-1177(94)90404-9.

Marx, Karl. *Critique of the Gotha Programme*. 1875. In *Marx/Engels Selected Works, Vol. Three*. Moscow: Progress Publishers, 1970. https://www.marxists.org/archive/marx/works/1875/gotha/ch01.htm.

Marx, Karl, and Frederick (Friedrich) Engels. *Manifesto of the Communist Party*. 1848. In *Marx/Engels Selected Works, Vol. One*, 98–137. Moscow: Progress Publishers, 1969. https://www.marxists.org/archive/marx/works/download/pdf/Manifesto.pdf.

McKay, Christopher P. "Does Mars Have Rights? An Approach to the Environmental Ethics of Planetary Engineering." In *Moral Expertise: Studies in Practical and Professional Ethics*, edited by Don MacNiven. New York: Routledge, 1990.

McKay, Christopher P. "On Terraforming Mars." *Extrapolation* 23, no. 4 (1982): 309–314. https://doi.org/10.3828%2Fextr.1982.23.4.309.

McKay, Christopher P., Owen B. Toon, and James F. Kasting. "Making Mars Habitable." *Nature* 352 (1991): 489–496. https://doi.org/10.1038/352489a0.

McKay, Tom. "Elon Musk: A New Life Awaits You in the Off-World Colonies—for a Price." *Gizmodo*, January 17, 2020. https://gizmodo.com/elon-musk-a-new-life-awaits-you-on-the-off-world-colon-1841071257.

McKie, Robin. "How Our Colonial Past Altered the Ecobalance of an Entire Planet." *The Guardian*, June 10, 2018. https://www.theguardian.com/science/2018/jun/10/colonialism-changed-earth-geology-claim-scientists.

Menegus, Bryan. "Exclusive: Amazon's Own Numbers Reveal Staggering Injury Rates at Staten Island Warehouse." *Gizmodo*, November 25, 2019. https://gizmodo.com/exclusive-amazons-own-numbers-reveal-staggering-injury-1840025032.

Miller, Melinda C. "Land and Racial Wealth Inequality." *American Economic Review: Papers & Proceedings* 101, no. 3 (2011): 371–376. https://doi.org/10.1257/aer.101.3.371.

Milligan, Tony. "Constrained Dissent and the Rights of Future Generations." In *Dissent, Revolution and Liberty Beyond Earth*, edited by Charles S. Cockell, 7–20. Cham, Switzerland: Springer, 2016.

Milligan, Tony. "Rawlsian Deliberation about Space Settlement." In *Human Governance Beyond Earth: Implications for Freedom*, edited by Charles S. Cockell, 9–22. Cham, Switzerland: Springer, 2015.

Mitchell, T. L. *Three Expeditions into the Interior of Eastern Australia*. Vol. 1. London: T. and W. Boone, 1839.

Montgomery, Laura. "Ep 03: Who Owns Mars?" Interview by Erika Nesvold. *Making New Worlds*, November 29, 2017. Podcast audio and transcript. https://makingnew worlds.com/2017/11/29/ep-03-who-owns-mars/.

Montgomery, Laura. "Ep 06: Who's in Charge?" Interview by Erika Nesvold. *Making New Worlds*, December 20, 2017. Podcast audio and transcript. https://makingnew worlds.com/2017/12/20/ep-06-whos-in-charge/.

Montgomery, Laura (Attorney and Sole Proprietor, Ground Based Space Matters). *Regulating Space: Innovation, Liberty, and International Obligations: Hearing before the Subcommittee on Space, Committee on Science, Space, and Technology, House of Representatives*. 115th Cong. (March 8, 2017). https://www.govinfo.gov/content/pkg/CHRG -115hhrg24671/html/CHRG-115hhrg24671.htm.

Morgan Stanley. *Investment Implications of the Final Frontier*. October 12, 2017. http:// www.fullertreacymoney.com/system/data/files/PDFs/2017/October/20th/msspace.pdf.

Musk, Elon. "Making Humans a Multi-Planetary Species." *New Space* 5, no. 2 (2017): 46–61. https://doi.org/10.1089/space.2017.29009.emu.

Nagin, Daniel S. "Deterrence in the Twenty-First Century." *Crime and Justice* 42 (2013): 199–263. https://doi.org/10.1086/670398.

NASA Orbital Debris Program Office. "Frequently Asked Questions." Accessed November 1, 2020. https://orbitaldebris.jsc.nasa.gov/faq/.

NASA. "NASA's Newest Astronaut Recruits to Conduct Research off the Earth, for the Earth and Deep Space Missions." Press release, June 7, 2017. https://www.nasa.gov /press-release/nasa-s-newest-astronaut-recruits-to-conduct-research-off-the-earth-for -the-earth-and.

National Space Society. "NSS Statement of Philosophy." Accessed March 29, 2020. https://space.nss.org/nss-statement-of-philosophy/.

Natoli, Jaime L., Deborah L. Ackerman, Suzanne McDermott, and Janice G. Edwards. "Prenatal Diagnosis of Down Syndrome: A Systematic Review of Termination Rates (1995–2011)." *Prenatal Diagnosis* 32, no. 2 (2012): 142–153. https://doi.org/10.1002/pd.2910.

Naval Photographic Center. "Trial by Fire: A Carrier Fights For Life." Produced 1973. Educational video, 18:43. https://www.youtube.com/watch?v=U6NnfRT_OZA.

Newell, Margaret. "Ep 03: Who Owns Mars?" Interview by Erika Nesvold. *Making New Worlds*, November 29, 2017. Podcast audio and transcript. https://makingnewworlds .com/2017/11/29/ep-03-who-owns-mars/.

Newell, Margaret. "Ep 04: Where's My Money?" Interview by Erika Nesvold. *Making New Worlds*, December 6, 2017. Podcast audio and transcript. https://makingnew worlds.com/2017/12/06/ep-04-wheres-my-money/.

Newell, Sarah. "Ep 04: Where's My Money?" Interview by Erika Nesvold. *Making New Worlds*, December 6, 2017. Podcast audio and transcript. https://makingnewworlds .com/2017/12/06/ep-04-wheres-my-money/.

Newman, Chris. "Ep 05: What If Someone Steals My Stuff?" Interview by Erika Nesvold. *Making New Worlds*, December 13, 2017. Podcast audio and transcript. https:// makingnewworlds.com/2017/12/13/ep-05-what-if-someone-steals-my-stuff/.

Newman, Chris. "Ep 06: Who's in Charge?" Interview by Erika Nesvold. *Making New Worlds*, December 20, 2017. Podcast audio and transcript. https://makingnewworlds .com/2017/12/20/ep-06-whos-in-charge/.

Newman, Chris. "Ep 08: Should We Make Mars More Like Earth?" Interview by Erika Nesvold. *Making New Worlds*, January 17, 2018. Podcast audio and transcript. https:// makingnewworlds.com/2018/01/17/ep-08-should-we-make-mars-more-like-earth -terraforming-and-environmental-conservation/.

Nguyen, Amanda, Erika Nesvold, Henry Hertzfeld, and Yuliya Panfil. "Law & Order, or Game of Thrones? The Legal Landscape of Space Exploration." Panel discussion at New America's How Will We Govern Ourselves in Space?, Washington, DC, July 10, 2019. http://opentranscripts.org/transcript/legal-landscape-space-exploration/.

Oberhaus, Daniel. "A Crashed Israeli Lunar Lander Spilled Tardigrades on the Moon." *Wired*, August 5, 2019. https://www.wired.com/story/a-crashed-israeli-lunar-lander -spilled-tardigrades-on-the-moon/.

Oman-Reagan, Michael P. "Queering Outer Space." Unpublished manuscript, submitted January 22, 2017, last modified July 2, 2018. SocArXiv, Open Science Framework. https://doi.org/10.31235/osf.io/mpyk6.

O'Neill, Gerard K. *The High Frontier: Human Colonies in Space*. New York: William Morrow, 1977.

Organ Procurement and Transplantation Network Ethics Committee. *Ethical Principles to Be Considered in the Allocation of Human Organs*. OPTN White Paper, 2010. https://optn.transplant.hrsa.gov/resources/ethics/ethical-principles-in-the-allocation -of-human-organs/.

Ostrom, Elinor. *Governing the Commons: The Evolution of Institutions for Collective Actions*. Cambridge: Cambridge University Press, 1990.

Ozalp, Mehmet. "Ep 11: Which Way Is Mecca?" Interview by Erika Nesvold. *Making New Worlds*, February 7, 2018. Podcast audio and transcript. https://makingnewworlds .com/2018/02/07/ep-11-which-way-is-mecca/.

Pandya, Shawna. "Ep 07: What If I Want to Have Kids?" Interview by Erika Nesvold. *Making New Worlds*, January 3, 2018. Podcast audio and transcript. https://makingnew worlds.com/2018/01/03/ep-07-what-if-i-want-to-have-kids/.

Parmitano, Luca. "EVA 23: Exploring the Frontier." *Luca blog*. European Space Agency, August 20, 2013. http://blogs.esa.int/luca-parmitano/2013/08/20/eva-23-exploring-the -frontier/.

Pascoe, Bruce. *Dark Emu: Aboriginal Australia and the Birth of Agriculture*. London: Scribe, 2018.

Pavard, Bibia. "The Right to Know? The Politics of Information about Contraception in France (1950s–'80s)." *Medical History* 63, no. 2 (2019): 173–188. https://dx.doi .org/10.1017%2Fmdh.2019.4.

Perry, David. "The Ruderman White Paper: On Media Coverage of the Murder of People with Disabilities by Their Caregivers." Ruderman Family Foundation, March 2017. https://rudermanfoundation.org/white_papers/media-coverage-of-the-murder-of -people-with-disabilities-by-their-caregivers/.

Persad, Govind, Alan Wertheimer, and Ezekiel J. Emanuel. "Principles for Allocation of Scarce Medical Interventions." *The Lancet* 373, no. 9661 (2009): 423–431. https://doi .org/10.1016/S0140-6736(09)60137-9.

Persson, Erik. "Citizens of Mars Ltd." In *Human Governance Beyond Earth: Implications for Freedom*, edited by Charles S. Cockell, 121–137. Cham, Switzerland: Springer, 2015.

Pilkington, Ed. "US Inmates Stage Nationwide Prison Labor Strike Over 'Modern Slavery.'" *The Guardian*, August 21, 2018. https://www.theguardian.com/us-news/2018 /aug/20/prison-labor-protest-america-jailhouse-lawyers-speak.

Plait, Phil. "The 50th Anniversary of Starfish Prime: The Nuke that Shook the World." *Discover*, July 9, 2012. https://www.discovermagazine.com/the-sciences/the -50th-anniversary-of-starfish-prime-the-nuke-that-shook-the-world.

Potter, Sean, ed. "Explorers Wanted: NASA to Hire More Artemis Generation Astronauts." NASA press release, February 11, 2020. https://www.nasa.gov/press-release /explorers-wanted-nasa-to-hire-more-artemis-generation-astronauts.

Potthast, Adam. "Alien Attacks, Hell Gerbils, and Assisted Dying: Arguments Against Saving Mere Humanity." *Futures* 110 (2019): 41–43. https://doi.org/10.1016/j .futures.2019.02.008.

Powelson, John P. *The Story of Land: A World History of Land Tenure and Agrarian Reform*. Cambridge, MA: Lincoln Institute of Land Policy, 1988.

Price, David A. *Love and Hate in Jamestown: John Smith, Pocahontas, and the Start of a New Nation*. Reprint ed. New York: Vintage, 2007.

Prochnau, William, and Laura Parker. "Trouble in Paradise." *Vanity Fair*, December 17, 2007. https://www.vanityfair.com/news/2008/01/pitcairn200801.

Quinones, Julian, and Arijeta Lajka. "'What Kind of Society Do You Want to Live In?':
Inside the Country Where Down Syndrome Is Disappearing." *CBS News*, August 14,
2017. https://www.cbsnews.com/news/down-syndrome-iceland/.

Race, Margaret. "Ep 09: What If There's Already Life on Mars?" Interview by Erika
Nesvold. *Making New Worlds*, January 24, 2018. Podcast audio and transcript. https://
makingnewworlds.com/2018/01/24/ep-09-what-if-theres-already-life-on-mars-planetary
-protection/.

Rand, Lisa Ruth. "Falling Cosmos: Nuclear Reentry and the Environmental History
of Earth Orbit." *Environmental History* 24, no. 1 (2019): 78–103. https://doi.org/10.1093
/envhis/emy125.

Rawls, John. *A Theory of Justice*. Cambridge, MA: Belknap Press, 1971.

Raworth, Kate. "A Safe and Just Space for Humanity." *Oxfam Discussion Papers* (2012).
https://www.oxfam.org/en/research/safe-and-just-space-humanity.

Riederer, Rachel. "Silicon Valley Says Space Mining Is Awesome and Will Change Life
on Earth. That's Only Half Right." *New Republic*, May 19, 2014. https://newrepublic.com
/article/117815/space-mining-will-not-solve-earths-conflict-over-natural-resources.

Robbins, Martin. "How Can Our Future Mars Colonies Be Free of Sexism and Racism?"
The Guardian, May 6, 2015. https://www.theguardian.com/science/the-lay-scientist/2015
/may/06/how-can-our-future-mars-colonies-be-free-of-sexism-and-racism.

Robertson, Adi. "SpaceX Wants to Be the Railroad of the Future." *The Verge*, September 27, 2016. https://www.theverge.com/2016/9/27/13080970/spacex-elon-musk-mars
-expedition-railroad-of-the-future.

Rogozov, L. I. "Self Operation." *Soviet Antarctic Expedition Information Bulletin* 4 (1964):
223–224. English translation. https://www.southpolestation.com/trivia/igy1/rogozov
.pdf.

Rolston, Holmes, III. "The Preservation of Natural Value in the Solar System." In *Beyond Spaceship Earth: Environmental Ethics and the Solar System*, edited by Eugene C.
Hargrove, 140–182. San Francisco: Sierra Club Books, 1986.

Romo, Vanessa. "California Bill Clears Path for Ex-Inmates to Become Firefighters."
NPR, September 11, 2020. https://www.npr.org/2020/09/11/912193742/california-bill-clears
-path-for-ex-inmates-to-become-firefighters.

Ronca, April E., Bernd Fritzsch, Laura L. Bruce, and Jeffrey R. Alberts. "Orbital Spaceflight During Pregnancy Shapes Mammalian Vestibular System." *Behavioral Neuroscience* 122, no. 1 (2008): 224–232. https://doi.org/10.1037/0735-7044.122.1.224.

Rosenberg, Paul. "What Happens to Dissent When Flight Is a Present Option?" In *Dissent, Revolution and Liberty Beyond Earth*, edited by Charles S. Cockell, 155–167. Cham,
Switzerland: Springer, 2016.

Ross, Michael L. *The Oil Curse: How Petroleum Wealth Shapes the Development of Nations*. Princeton, NJ: Princeton University Press, 2012.

Ross, Michael L. "What Have We Learned about the Resource Curse?" *Annual Review of Political Science* 18 (2015): 239–259. https://doi.org/10.1146/annurev-polisci-052213-040359.

Roussos, Panos, Stella G. Giakoumaki, and Panos Bitsios. "Cognitive and Emotional Processing in High Novelty Seeking Associated with the L-DRD4 Genotype." *Neuropsychologia* 47, no. 7 (2009): 1654–1659. https://doi.org/10.1016/j.neuropsychologia.2009.02.005.

Sagan, Carl. *Cosmos*. New York: Random House, 1980.

Sagan, Carl. *Cosmos: A Personal Voyage*. Episode 7, "The Backbone of Night." Aired November 9, 1980.

Sagan, Carl. *Pale Blue Dot: A Vision of the Human Future in Space*. New York: Random House, 1994.

Sagan, Carl. "The Planet Venus." *Science* 133, no. 3456 (1961): 849–858. https://doi.org/10.1126%2Fscience.133.3456.849.

Sagan, Carl. "Planetary Engineering on Mars." *Icarus* 20, no. 4 (1973): 513–514. https://doi.org/10.1016%2F0019-1035%2873%2990026-2.

Salmeri, Antonino. "No, Mars Is Not a Free Planet, No Matter What SpaceX Says." *SpaceNews*, December 5, 2020. https://spacenews.com/op-ed-no-mars-is-not-a-free-planet-no-matter-what-spacex-says/.

Sanger, David E. "Soviets Send First Japanese, a Journalist, into Space." *New York Times*, December 3, 1990. https://nyti.ms/2O217Nh.

Sapp, Walter J., Delbert E. Philpott, Carol S. Williams, Katharine Kato, Joann Stevenson, M. Vasquez, and L. V. Serova. "Effects of Spaceflight on the Spermatogonial Population of Rat Seminiferous Epithelium." *FASEB Journal* 4, no. 1 (1990): 101–104. https://doi.org/10.1096/fasebj.4.1.2295370.

Scharmen, Fred. *Space Forces: A Critical History of Life in Outer Space*. London: Verso, 2021.

Schröder, K.-P., and Robert Connon Smith. "Distant Future of the Sun and Earth Revisited." *Monthly Notices of the Royal Astronomical Society* 386, no. 1 (2008): 155–163. https://doi.org/10.1111/j.1365-2966.2008.13022.x.

Schwartz, James S. J. "Fairness as a Moral Grounding for Space Policy." In *The Meaning of Liberty Beyond Earth*, edited by Charles S. Cockell, 69–89. Cham, Switzerland: Springer, 2015.

Schwartz, James S. J. "Ep 08: Should We Make Mars More Like Earth?" Interview by Erika Nesvold. *Making New Worlds*, January 17, 2018. Podcast audio and transcript. https://makingnewworlds.com/2018/01/17/ep-08-should-we-make-mars-more-like-earth-terraforming-and-environmental-conservation/.

Schwartz, James S. J. "Myth-Free Space Advocacy Part II: The Myth of the Space Frontier." *Astropolitics* 15, no. 2 (2017): 167–184. http://doi.org/10.1080/14777622.2017.1339255.

Schwartz, James S. J., and Tony Milligan. "'Space Ethics' According to Space Ethicists." *Space Review*, February 1, 2021. https://www.thespacereview.com/article/4117/1.

Scientific American Editors. "How to Reinvent Policing." *Scientific American* 323, no. 3, 8 (2020). https://www.scientificamerican.com/article/three-ways-to-fix-toxic-policing/.

Shanahan, Jesse. "Ep 02: Who Gets to Go?" Interview by Erika Nesvold. *Making New Worlds*, November 22, 2017. Podcast audio and transcript. https://makingnewworlds.com/2017/11/22/episode-2-who-gets-to-go/.

Shayler, David. *Around the World in 84 Days: The Authorized Biography of Skylab Astronaut Jerry Carr*. Burlington, ON: Apogee, 2008.

Sherman, W. T. *Special Field Orders, No. 15*. Headquarters Military Division of the Mississippi, January 16, 1865. In *The Wartime Genesis of Free Labor: The Lower South*, edited by Ira Berlin, Thavolia Glymph, Steven F. Miller, Joseph P. Reidy, Leslie S. Rowland, and Julie Saville, 338–340. Cambridge, NY: Cambridge University Press, 2012. http://www.freedmen.umd.edu/sfo15.htm.

Siddiqi, Asif A. "Competing Technologies, National(ist) Narratives, and Universal Claims: Toward a Global History of Space Exploration." *Technology and Culture* 51, no. 2 (2010): 425–443. https://doi.org/10.1353/tech.0.0459.

Singh, Kanishka. "US Labor Judge Rules that Tesla Broke Labor Law." *Reuters*, September 27, 2019. https://www.reuters.com/article/us-tesla-labor/u-s-labor-judge-rules-that-tesla-broke-labor-law-idUSKBN1WD003.

Slave Voyages. "Trans-Atlantic Slave Trade Database." Accessed February 1, 2020. https://www.slavevoyages.org/assessment/estimates.

Smith, Cameron M. "Estimation of a Genetically Viable Population for Multigenerational Interstellar Voyaging: Review and Data for Project Hyperion." *Acta Astronautica* 97 (2014): 16–29. https://doi.org/10.1016/j.actaastro.2013.12.013.

Smith, Kelly C. "*Homo reductio*: Eco-Nihilism and Human Colonization of Other Worlds." *Futures* 110 (2019): 31–34. https://doi.org/10.1016/j.futures.2019.02.005.

Smith, Kelly C. "Ep 09: What If There's Already Life on Mars?" Interview by Erika Nesvold. *Making New Worlds*, January 24, 2018. Podcast audio and transcript. https://makingnewworlds.com/2018/01/24/ep-09-what-if-theres-already-life-on-mars-planetary-protection/.

Smits, David D. "The Frontier Army and the Destruction of the Buffalo: 1865–1883." *Western Historical Quarterly* 25, no. 3 (1994): 312–338. https://doi.org/10.2307/971110.

Sparrow, Robert. "The Ethics of Terraforming." *Environmental Ethics* 21, no. 3 (1999): 227–245. https://doi.org/10.5840/enviroethics199921315.

Starlink. "Starlink Pre-Order Agreement." Accessed April 7, 2021. https://www.starlink.com/legal/documents/DOC-1020-91087-64?regionCode=US.

Stern, Alexandra Minna. "Forced Sterilization Policies in the US Targeted Minorities and Those with Disabilities—and Lasted into the 21st Century." *The Conversation*,

August 26, 2020. https://theconversation.com/forced-sterilization-policies-in-the-us
-targeted-minorities-and-those-with-disabilities-and-lasted-into-the-21st-century
-143144.

Stewart, Henry P. "The Impact of the USS *Forrestal*'s 1967 Fire on United States Navy Shipboard Damage Control." Master's thesis, US Army Command and General Staff College, 2004. https://apps.dtic.mil/sti/pdfs/ADA429103.pdf.

Stewart, Will. "Antarctic Scientist 'Stabs Colleague Who Kept Telling Him the Endings of Books He Was Reading on Remote Research Station.'" *The Sun*, October 30, 2018. https://www.thesun.co.uk/news/7615571/antarctic-scientist-stabs-colleague-who
-kept-telling-him-endings-of-books-he-was-reading/.

Strang, Heather, Lawrence W. Sherman, Evan Mayo-Wilson, Daniel Woods, and Barak Ariel. "Restorative Justice Conferencing (RJC) Using Face-to-Face Meetings of Offenders and Victims: Effects on Offender Recidivism and Victim Satisfaction. A Systematic Review." *Campbell Systematic Reviews* 9, no. 1 (2013). https://doi.org/10.4073/csr.2013.12.

Taylor, Chris. "'I'm the First Space Pirate!' How Tardigrades Were Secretly Smuggled to the Moon." *Mashable*, August 8, 2019. https://mashable.com/article/smuggled
-moon-tardigrade/.

Thiere, Adam. *Permissionless Innovation: The Continuing Case for Comprehensive Technological Freedom*. Arlington, VA: Mercatus Center at George Mason University, 2016.

Thomas, Jessica E., Gary R. Carvalho, James Haile, Nicolas J. Rawlence, Michael D. Martin, Simon Y. W. Ho, Arnór Þ Sigfússon, Vigfús A. Jósefsson, Morten Frederiksen, Jannie F. Linnebjerg, Jose A. Samaniego Castruita, Jonas Niemann, Mikkel-Holger S. Sinding, Marcela Sandoval-Velasco, André E. R. Soares, Robert Lacy, Christina Barilaro, Julia Best, Dirk Brandis, Chiara Cavallo, Mikelo Elorza, Kimball L. Garrett, Maaike Groot, Friederike Johansson, Jan T. Lifjeld, Göran Nilson, Dale Serjeanston, Paul Sweet, Errol Fuller, Anne Karin Hufthammer, Morten Melgaard, Jon Fjeldså, Beth Shapiro, Michael Hofreiter, John R. Stewart, M. Thomas P. Gilbert, and Michael Knapp. "Demographic Reconstruction from Ancient DNA Supports Rapid Extinction of the Great Auk." *eLife* 8 (2019): e47509. https://doi.org/10.7554%2FeLife.47509.

Tronchetti, Fabio. "The Space Resource Exploration and Utilization Act: A Move Forward or a Step Back?" *Space Policy* 34 (2015): 6–10. https://doi.org/10.1016/j.spacepol
.2015.08.001.

Turner, Brad, dir. *Battlestar Galactica*. Season 1, episode 8, "Flesh and Bone." Aired February 25, 2005, on Sci Fi.

UN Food and Agriculture Organization. *The State of World Fisheries and Aquaculture 2020: Sustainability in Action*. Rome: FAO, 2020. https://doi.org/10.4060/ca9229en.

UN General Assembly. "Agreement Governing the Activities of States on the Moon and Other Celestial Bodies." RES 34/68. 1979. https://www.unoosa.org/oosa/en/ourwork
/spacelaw/treaties/moon-agreement.html.

UN General Assembly. "Treaty on Principles Governing the Activities of States in the Exploration and Use of Outer Space, Including the Moon and Other Celestial Bodies." 2222 (XXI). 1967. http://www.unoosa.org/oosa/en/ourwork/spacelaw/treaties/outer spacetreaty.html.

UN General Assembly. "Universal Declaration of Humans Rights." 217 (III) A. Paris, 1948. http://www.un.org/en/universal-declaration-human-rights/.

US Census Bureau. "Enumeration of Persons in the Several Districts of the United States." December 8, 1801. https://www2.census.gov/library/publications/decennial /1800/1800-returns.pdf.

US Census Bureau. "Top 10 Most Populous Countries (July 1, 2019)." Accessed January 7, 2020. https://www.census.gov/popclock/print.php?component=counter.

US Commercial Space Launch Competitiveness Act. H.R. 2262. 114th Cong. (2015). https://www.congress.gov/bill/114th-congress/house-bill/2262.

US National Center for Health Workforce Analysis. "Sex, Race, and Ethnic Diversity of US Health Occupations (2011–2015)." Washington, DC: US Department of Health and Human Services, August 2017. https://bhw.hrsa.gov/sites/default/files/bureau -health-workforce/data-research/diversity-us-health-occupations.pdf.

Valentine, David, Valerie A. Olson, and Debbora Battaglia. "Encountering the Future: Anthropology and Outer Space." *Anthropology News* 50, no. 9 (2009): 11–15. https://doi .org/10.1111/j.1556-3502.2009.50911.x.

Veatch, Robert M. "Disaster Preparedness and Triage: Justice and the Common Good." *Mount Sinai Journal of Medicine* 72, no. 4 (2005): 236–241. https://www.researchgate .net/publication/7724070_Disaster_Preparedness_and_Triage_Justice_and_the_Com mon_Good.

Vincent, Alice. "Ombudspersons for Future Generations: Bringing Intergenerational Justice into the Heart of Policymaking." *UN Chronicle* 49, no. 2 (2012): 66–68. https:// doi.org/10.18356/2c3c1e22-en.

Vitale, Alex S. *The End of Policing*. London: Verso, 2017.

Wakayama, Sayaka, Yuko Kamada, Kaori Yamanaka, Takashi Kohda, Hiromi Suzuki, Toru Shimazu, Motoki N. Tada, Ikuko Osada, Aiko Nagamatsu, Satoshi Kamimura, Hiroaki Nagatomo, Eiji Mizutani, Fumitoshi Ishino, Sachiko Yano, and Teruhiko Wakayama. "Healthy Offspring from Freeze-Dried Mouse Spermatozoa Held on the International Space Station for 9 Months." *PNAS* 114, no. 23 (2017): 5988–5993. https:// doi.org/10.1073/pnas.1701425114.

Wakayama, Sayaka, Yumi Kawahara, Chong Li, Kazuo Yamagata, Louis Yuge, and Teruhiko Wakayama. "Detrimental Effects of Microgravity on Mouse Preimplantation Development In Vitro." *PLoS ONE* 4, no. 8 (2009): e6753. https://doi.org/10.1371/journal .pone.0006753.

Wall, Mike. "Foam 'Spider Webs' from Tiny Satellites Could Help Clean Up Space Junk." *Space.com*, June 23, 2020. https://www.space.com/space-junk-cleanup-foam -satellite-technology.html.

Waltemathe, Michael. "Ep 11: Which Way Is Mecca?" Interview by Erika Nesvold. *Making New Worlds*, February 7, 2018. Podcast audio and transcript. https://makingnew worlds.com/2018/02/07/ep-11-which-way-is-mecca/.

Ward, Peter. *The Consequential Frontier: Challenging the Privatization of Space*. Brooklyn: Melville House, 2019.

Wei, Will. "Peter Diamandis: The First Trillionaire Is Going to Be Made in Space." *Business Insider*, March 2, 2015. https://www.businessinsider.com/peter-diamandis -space-trillionaire-entrepreneur-2015-2.

Weibel, Deana L. "The Overview Effect and the Ultraview Effect: How Extreme Experiences in/of Outer Space Influence Religious Beliefs in Astronauts." *Religions* 11, no. 8 (2020): 418–434. https://www.doi.org/10.3390/rel11080418.

Weinzierl, Matthew. "Space, the Final Economic Frontier." *Journal of Economic Perspectives* 32, no. 2 (2018): 173–192. https://www.jstor.org/stable/26409430.

White, Frank. *The Overview Effect: Space Exploration and Human Evolution*. Boston: Houghton-Mifflin, 1987.

White, Kami. "Virgin Galactic Welcomes Two New Pilots." Virgin Galactic, October 27, 2020. https://www.virgin.com/about-virgin/latest/virgin-galactic-welcomes-two-new -pilots.

Wilcove, David S., David Rothstein, Jason Dubow, Ali Phillips, and Elizabeth Losos. "Quantifying Threats to Imperiled Species in the United States." *BioScience* 48, no. 8 (1998): 607–615. https://doi.org/10.2307/1313420.

Williams, Casey. "The State Locked Them Up. Now They're North Carolina's Elite Firefighting Force." *Scalawag Magazine*, May 24, 2018. https://www.scalawagmagazine .org/2018/05/the-state-locked-them-up-now-theyre-north-carolinas-elite-firefighting -force/.

Wohlforth, Charles, and Amanda R. Hendrix. *Beyond Earth: Our Path to a New Home in the Planets*. New York: Knopf Doubleday Publishing Group, 2016.

Wong, Julia Carrie. "Tesla Factory Workers Reveal Pain, Injury and Stress: 'Everything Feels Like the Future but Us.'" *The Guardian*, May 18, 2017. https://www.theguardian .com/technology/2017/may/18/tesla-workers-factory-conditions-elon-musk.

World Bank Group. "Population, Female (% of Total Population)." Accessed January 7, 2020. https://data.worldbank.org/indicator/sp.pop.totl.fe.zs.

Wright, Ronald. *A Short History of Progress*. Toronto: House of Anansi, 2004.

Zapata, Edgar. "An Assessment of Cost Improvements in the NASA COTS-CSR Program and Implications for Future NASA Missions." NASA (2018). https://ntrs.nasa.gov /archive/nasa/casi.ntrs.nasa.gov/20170008895.pdf.

Zevallos, Zuleyka. "Ep 01: Why Are We Going?" Interview by Erika Nesvold. *Making New Worlds*, November 15, 2017. Podcast audio and transcript. https://makingnew worlds.com/2017/11/15/episode-1-why-are-we-going/.

Zubrin, Robert. *The Case for Mars: The Plan to Settle the Red Planet and Why We Must.* New York: Touchstone, 1996.

Zubrin, Robert. "The Tardigrades-on-the-Moon Affair." *National Review*, August 31, 2019. https://www.nationalreview.com/2019/08/planetary-protection-rules-hamper-space -exploration/.

Zubrin, Robert M., and Christopher P. McKay. "Technological Requirements for Terra-forming Mars." *AIAA-03–2005* (1993). https://doi.org/10.2514/6.1993-2005.

INDEX

Forward contamination, 98–99
  prevention of, 99–100
Free-riders, 57
French, Robert, 109–110
Friedrich, Otto, 137
Frontier image, xii, 10
  critique of, 10–12, 23–27
  and justice, 140–141
  O'Neill and, xii–xiii
Future generations, xiv
  and culture, 214–215
  and disability, 44
  and government, 151, 162
  and property rights, 66
  and resource exploitation, 76–77, 82–
    87, 89–91
  and sustainability, 87–89
Fyodorov, Nikolai, 6

Gabaccia, Donna, 27, 32
Gagarin, Yuri, 40
Gandhi, Indira, 177
Gandhi, Rajiv, 177
Garriott, Richard, 231n6
Genetics
  diversity and, 46–47, 181
  evolutionary argument and, 8–9
  underpopulation and, 180
Geneva Convention, 195
Geostationary orbit, 82
Glenn, Alice Qannik, 205–208, 214
Godwin, Tom, 140
Gonzales-Day, Ken, 141
Good life, visions for, xiii, 89, 145
Gore, Al, 86
Gorman, Alice, 70
Government, 147–148
  historical models and, 155–159
  local, 150–153
  and property rights, 59, 62
  and selection criteria, 37
Gravity
  and children, 184
  and reproduction, 171
  and weapons, 160

Gravity (film), 79
Greece, ancient, 163
Guardians
  for extraterrestrial life, 109
  for Whanganui River, 70

Habitat destruction, 96–97
Haitian Revolution, 147, 158–159
Hale, Edward Everett, xi
Hardin, Garrett, 31, 57
Harris, John, 193
Hartmann, William K., 15
Hawking, Stephen, 14
Headrights, 61
Health care. *See also* Triage
  emergency preparedness and, 190–193
  scarcity of resources and, 188, 193
Health care workers, as care priority,
    197–198
Herbert, Nick, 141
Hertzfeld, Henry, 59–60
Hierarchies, 149
High Frontier, The (O'Neill), xii–xiii, 10
Historical models
  critique of, 23–27
  and economic issues, 116–118
  and environment, 96–97
  further study on, 218–219
  and government, 155–159
  Jamestown settlement as, 48–49
  and land distribution, 60–63
  and law enforcement, 131
  and prison, 137
  and resource exploitation, 75–77
  and triage, 187, 194
Hoffman, Jeff, 213
Holthaus, Eric, 89
Homelessness, 122
Horneck, Gerda, 89
Humanities, value of, viii, 221
Human nature
  arguments from, 7–8, 10
  and hierarchies, 149
  persistence of, 143–144
  and war, 159–161

Leadership, 147–148. *See also*
   Government
   and judging, 133
   and policing, 131
Learning skills, as selection criterion,
   42
Legal issues, 129–130
   current system, 153–156, 163
   and dissent, 153
   further learning on, 219
   and non-human entities, 69–70
   and population, 177–179
   and property rights, 59–60, 63–66
   and resources, 84, 89
Leopold, Aldo, 95, 106
Liberty, 162–163
Life, extraterrestrial
   contamination of Earth and, 98
   environmental precautions and,
      97
   terraforming and, 103–104
Lifeboat scenario, 30–33
Life support
   overpopulation and, 176
   and poverty, 122–123
   and tyranny, 152
   war and, 160–161
Little Badger, Darcie, 25
Local government, 150–153
Locke, John, 60
Logsdon, John, 101
Lottery system
   and health care, 193, 196
   and land distribution, 58
   and settler selection, 39, 41
Lupisella, Mark, 101
Lynching, 141
Lyons, Oren, 86

Magellan, Ferdinand, 27
Malingering, 129
Mallory, George, 22
Manifest Destiny, 6–7
   critique of, 12
Mann, Charles C., 69
Maritime laws, 154–155

Mars
   Starlink and, 155–156
   terraforming, 103
Marx, Karl, 67
McKay, Christopher, 102, 104
Medicine. *See also* Triage
   emergency preparedness and,
      190–193
   scarcity of resources and, 188, 193
Mental health
   care, 191–192
   space environment and, 207, 213
Mercury, 99
Microgravity manufacturing, 74
Military
   and hierarchy, 150–151
   recruitment from, 22
   and selection criteria, 43
   and triage, 187, 194, 197
Milligan, Tony, 90, 151, 153, 220
Mining companies, 74
   costs of, 76
   and overuse, 82–85
   and pollution, 84–85
   and profit, 114
   and property rights, 65–66
   regulation of, 90
Mission value, and triage, 197
Montgomery, Laura, 56, 65, 154, 157
Montoya, Courtney, 135
Moon Agreement, 65, 89
Moon Express, 64
Moore, William, xi
Motivations for space settlement,
   21–22
   destiny as, 23–27
   and economic status, 124
   legacy of, 33–34
   and methods, 22–23
   progress as, 27–30
   religion and, 209–210
   survival as, 30–33
Movement, freedom of, 38, 163
Muir, John, 106
Musk, Elon, 14, 29, 48, 115, 125
Muslims, 210–212, 214–215

Private space industry
  and economic issues, 114
  and government, 153–156
  and labor issues, 124–125
  and medical rationing, 199
  and planetary protection, 102
  and preparedness, vii–viii
  and regulation, 90
  and selection criteria, 37
Privatization, of commons, 57
Profit, 154
  issues with, 27–30
  prison labor and, 138
Progress
  as argument, 9–13
  as motivation, 27–30
Property, nature of, 59–60
Property rights, 53–54
  alternatives to, 66–70
  benefits of, 55–58
  importance of, 54–55
  Marx on, 67
  sociological components of, 57–59
  theft and, 127–128
Public collaboration, xiii–xiv
  and triage, 200–201
Punishment, prison and, 134

Quality-adjusted life year (QALY), 195
  and triage, 195–196
Questions for preparedness, viii, 18–19
  on environment, 95–96
  on government, 151

Race, Margaret, 99–100
Race issues
  and justice, 141
  and police, 131–132
  and prison, 139–140
  and prison labor, 138
  and reproduction, 178–179
Railroads, and labor, 113, 119–120
Rand, Lisa Ruth, 220
Rationing, medical, 199
Rawls, John, 86, 151–152
Raworth, Kate, 88

Readiness, nature of, 17
Recidivism, reduction of, 134, 143
Redemptioners, 29–30
Refugees, 21, 31–33
Regulation
  of commons, 57
  vs. labor exploitation, 117–118
  and planetary protection, 101–102
  of resource extraction, 90
Rehabilitation, prison and, 134
Relationships with Earth, 148–149
  historical models and, 156–159
  remote medical consultation, 192
Religion, 203
  changes in, 212–213
  and destiny argument, 6–7
  practices, in space, 208–212
  recommendations for, 214–215
  and species value, 95
Representation, and selection criteria,
    40–41
Reproduction
  encouraging, 179–182
  ethics of, 170–171, 173–175, 177–178, 185
  and selection criteria, 45–46
  in space, issues with, 171–173
Rescue, rule of, 194
Research
  and preparedness, 218–221
  on reproduction in space, 172–173
Resource curse, 75–76
Restorative justice, 139–143, 219
Revolution, 156–159
Richards, Bob, 65–66
Rights
  of children, 182–184
  and population issues, 170
Robbins, Martin, 12
Rogozov, Leonid, 191
Role models, xi
  and advisability of space settlement, 4
  critique of, 23–27
Rolston, Holmes, III, 107
Rosenberg, Paul, 163
Ross, Michael L., 76
Rule of rescue, 194